Academic and Professional Writing
in an Age of Accountability

Academic and Professional Writing in an Age of Accountability

<hr>

Edited by Shirley Wilson Logan and Wayne H. Slater

Afterword by Jessica Enoch and Scott Wible

Southern Illinois University Press • *Carbondale*

Southern Illinois University Press
www.siupress.com

Copyright © 2018 by the Board of Trustees,
Southern Illinois University

Chapter 1 copyright © 2018 by Robert Coogan, Jane Donawerth, and Molly Scanlon; chapter 2 copyright © 2018 by Jeanne Fahnestock; chapter 3 copyright © 2018 by Kelly A. Ritter; chapter 4 copyright © 2018 by Teresa M. Redd; chapter 6 copyright 2018 © by Cheryl Glenn; chapter 7 copyright © 2018 by James Paul Gee; chapter 8 copyright © 2018 by Suresh Canagarajah; chapter 9 copyright © by Staci M. Perryman-Clark; chapter 10 copyright © 2018 by Charles Bazerman; chapter 11 copyright © 2018 by James A. Herrick; chapter 12 copyright © 2018 by Douglas Eyman; chapter 13 copyright © 2018 by Anne Frances Wysocki; chapter 14 copyright © 2018 by Kathleen Blake Yancey; chapter 15 copyright © by Douglas D. Hesse

Publication of this book has been partially funded by a
subvention from the University of Maryland.

21 20 19 18   4 3 2 1

Library of Congress Cataloging-in-Publication Data
Names: Logan, Shirley W. (Shirley Wilson), [date] editor. | Slater, Wayne Harvard, editor. | Enoch, Jessica, writer of afterword. | Wible, Scott, [date] writer of afterword.
Title: Academic and professional writing in an age of accountability / edited by Shirley Wilson Logan and Wayne H. Slater ; afterword by Jessica Enoch and Scott Wible.
Description: Carbondale : Southern Illinois University Press, [2018] | Includes bibliographical references and index.
Identifiers: LCCN 2018017632 | ISBN 9780809336913 (paperback : alk. paper) | ISBN 9780809336920 (e-book)
Subjects: LCSH: Academic writing—Study and teaching (Higher)—United States. | Business writing—Study and teaching (Higher)—United States. | Technical writing—Study and teaching (Higher)—United States. | Rhetoric—Study and teaching (Higher)—United States.
Classification: LCC P301.5.A27 A23 2018 | DDC 808/.0420711—dc23
LC record available at https://lccn.loc.gov/2018017632

Printed on recycled paper. ♻

Shirley Logan dedicates this book to
Chandler and Harper Logan, and Jacob and Grayson Allen

Wayne Slater dedicates this book to
Ralph Harvard Slater and Mary Brackenridge Slater

Contents

Contents

SECTION IV. DIGITAL PERSPECTIVES

SECTION V. TWO PERSPECTIVES ON ASSESSMENT

Acknowledgments

The origins of this collection can be traced to the 2014 Maryland Conference on Academic and Professional Writing, held at the University of Maryland. With the generous support of the Office of the Provost, campus schools and colleges, and the University System of Maryland, the Campus Writing Board was able to assemble some of the leading writing and rhetoric scholars, writing program administrators, linguists, and journal editors to participate in a stimulating two-day meeting about the history, current state, and future direction of academic and professional writing. To all those involved in this event, we owe a debt of gratitude, particularly former associate provost and dean of undergraduate studies Donna B. Hamilton, who was an early and sustaining champion of the conference. Most of the conference speakers agreed to contribute to the volume, and we invited others to join them. We thank all of these scholars for their enthusiastic response and their thought-provoking essays.

We thank the Department of English and the Department of Teaching and Learning, Policy and Leadership for providing subvention support. We wish to thank editors Karl Kageff and Kristine Priddy at Southern Illinois University Press for their guidance. We are especially grateful to Robert Brown for his careful copyediting. Our family members have also been especially generous with their support.

Academic and Professional Writing
in an Age of Accountability

Introduction

Wayne H. Slater and Shirley Wilson Logan

In the United States, as David Bartholomae explains, university writing courses were already a century old by 1900, with writing instruction and the writing of papers having been a part of the university curriculum throughout the nineteenth century, informed by the work of Blair's *Lectures on Rhetoric and Belles Lettres* (1783), Newman's *A Practical System of Rhetoric* (1827), Parker's *Aids to English Composition* (1844), Boyd's *Elements of Rhetoric and Literary Criticism* (1844), and Quackenbos's *Advanced Course of Rhetoric and English Composition* (1855, 1950). In many ways, these writing courses, grounded in the first 2,500 years of Western literacy, represented a democratic ideal that effective writing empowers everyone in a free society (Bartholomae; Berlin; Kitzhaber; J. Murphy; Varnum; Wozniak).

As Bartholomae notes (1950), in the early 1900s, with the reconceptualization of the American university imitating the German model in its emphasis on research and the creation of departments, English departments were still committed to the teaching of writing. During this period, the issues raised in the professional literature focused on who staffs composition courses and how they should be prepared, the content of composition courses, evidence that composition courses improve students' writing, and the investment of money and time in composition courses relative to their value to students.

From 1901–1903, the Pedagogical Section of the Modern Language Association (MLA) contributed significantly to the discussion of these topics with Fred Newton Scott, University of Michigan English professor, providing leadership by encouraging the MLA Pedagogical Section to publish a series of reports in the *Publications of the Modern Language Association* (Applebee; Bartholomae; Brereton). Scott would later serve as president of the MLA in 1907 and as president of the National Council of Teachers of English (NCTE) in 1911. In 1901 a Pedagogical Section committee sent out a questionnaire focused on rhetoric as a graduate study concentration. A majority of the respondents supported the advanced study of rhetoric, with some questioning the goal: Was it to improve the writing of graduate students or to engage in new areas of research? A report concluded that the term *rhetoric* should be more comprehensive by including its use in the teaching of composition and as a theoretical framework for disciplined inquiry by graduate students (Bartholomae 1951).

In 1902 the Pedagogical Section committee surveyed MLA members about the worth of time and effort allocated to the teaching of composition and to ask whether student writing might be equally improved with readings in great literature. Overall, members supported composition instruction for students, with a number of them asserting that student writing improved as a result of taking a composition course (Bartholomae 1951). In the last of the reports, published in 1903, the Pedagogical Section surveyed departments on two issues: first, whether the goal of composition was to teach students to write clearly using accepted standard usage; second, whether composition was an academically respectable domain in the study of English or one that was repellently applied and remedial in function. The responses to the 1903 survey supported the view that composition is academically respectable, with student writing in composition classes being taken seriously as writing and not just as an exercise for learning standard usage. Interestingly, as English departments evolved into literature departments with growing enrollments, labor-intensive composition classes became marginalized and viewed with disdain, concurrently with their increasing distance from the influence of the MLA. This trend continued until the creation of the Conference on College Composition and Communication, or CCCC (Bartholomae 1951).

In 1949 CCCC met for the first time under the auspices of the NCTE. CCCC sponsored an annual meeting and a journal. As the organization focused its agenda, it published a bibliography and created awards for public recognition, established special interest groups, and created a community for scholars and teachers to cultivate a professional identity for those in composition (Applebee 160). In effect, CCCC addressed professional needs not provided by the MLA and NCTE. As Bartholomae notes, members of CCCC in the 1950s and 1960s provided the foundation for the development in the 1970s of composition as a defined field of inquiry with both research and pedagogical agendas, debates, and newly developed undergraduate and graduate programs in academic and professional writing and PhD programs in composition studies (1952). As in the early 1900s, many of the topics addressed at CCCC conferences today are remarkably similar: staffing and preparation for composition instruction, the content of composition courses, evidence that composition courses improve students' writing, and the investment of money and time in composition courses relative to their value to students.

Given the current state of composition, with many of its core issues essentially unchanged today, we are currently experiencing an age of increased accountability fueled by at least three curricular and accountability initiatives: the Common Core State Standards Initiative, the National Assessment of Educational Progress (NAEP), and the ACT College and Career Readiness campaign. An article in the June 9, 2016, *Washington Post*, "Common Core Isn't Preparing Students Very Well for College or Career, New Report Says," resurrects persisting worries about the efficacy of imposed standards (Strauss). It suggests that we have not yet developed the proper balance between meaningful teaching and effective assessment, a concern explored by Kathleen Blake Yancey in section V. The *Post* article describes a disconnect between the Common Core Standards and the expectations of college writing instructors with respect to narrative, persuasive, and informational genres of writing. At the center of this controversy (about which see, also, T. Murphy; Rich; Schneider), the supporters of the Common Core Standards claim, are a set of high-quality K–12 academic standards in English language arts and mathematics established to ensure that all students graduate from high school with the knowledge and skills necessary to succeed in college, career, and life, regardless of where they live. As of this writing, forty-two states, the District of Columbia, four territories, and the Department of Defense Education Activity have voluntarily adopted and are implementing the Common Core Standards (Common Core State Standards Initiative).

In concert with the Common Core Standards, the NAEP and the ACT College and Career Readiness campaign further contribute to ever-increasing expectations for more accountability and for curricula that ensure college and career readiness. The NAEP is the largest continuing and nationally representative assessment of what U.S. students know and are able to do in core K–12 subjects (National Assessment of Educational Progress). The NAEP is a congressionally mandated initiative administered by the National Center for Education Statistics within the Institute of Education Sciences of the U.S. Department of Education. The ACT, originally an abbreviation of the American College Test, is a standardized test for high school achievement and college admissions in the United States produced by ACT, Inc. The ACT's 2015 annual report on the progress of the most recent high school graduating class relative to college readiness synthesizes data from 1,924,436 students, or 59 percent of the 2015 graduating class (ACT, *Condition* 3). The percentage of graduates meeting the ACT's college readiness benchmarks in English has remained relatively steady. Beginning in 2010, the percentage of ACT-tested high school graduates meeting the benchmarks for English was 66 percent; in 2011 it remained 66 percent; in 2012 it was 67 percent; and in 2013, 2014, and 2015 it held at 64 percent (4). Taken together, the Common Core Standards,

NAEP, and ACT capture current critical trends in curricula and accountability that challenge the effectiveness of literacy development in the United States (ACT, *ACT National*). Their implementation and contested accountability expectations, for better or worse, support the need to improve college readiness in English and provide additional support for having academic and professional writing programs in colleges and universities.

The Campus Writing Board at the University of Maryland

The perennial concerns facing college writing programs today are remarkably similar to those articulated in the early twentieth century: the right selection and preparation of instructional staff, the proper content for writing courses, and evidence that writing instruction actually improves student writing. More recently, however, programs are concerned with the impact of the Common Core State Standards Initiative and increased emphasis on accountability. In response, in 2010 the University of Maryland's College of Arts and Humanities created the Campus Writing Board, a committee of faculty members from across the university who are charged with developing and supporting composition at Maryland. The faculty members discussed issues critical to the strategic and effective delivery of writing instruction. At the urging of the Campus Writing Board, we initiated meetings with every college and school head to discuss issues relevant to student writing at research universities and to obtain financial support for a proposed 2014 Maryland Conference on Academic and Professional Writing. We identified five topic domains as grounding our discussions: programmatic perspectives, rhetorical perspectives, language perspectives, digital perspectives, and assessment perspectives. Together these domains guided the program for the 2014 Maryland Conference on Academic and Professional Writing, which took place on October 10 and 11.

Emanating from the 2014 Maryland Conference, and sponsored by the University of Maryland's Campus Writing Board, this edited volume has emerged from five central questions: (1) What are the current theoretical frameworks that inform academic and professional writing? (2) What does research tell us about the range and effectiveness of academic and professional writing programs? (3) What do we know about current theory-based, research-driven best practices in academic and professional writing? (4) What are current directions in evaluating the effectiveness of academic and professional writing programs? (5) What are the future directions for research and best practices in academic and professional writing in an age of accountability? These five were some of the questions being asked by parents, teachers, education specialists, administrators, employers, and the Campus Writing Board. We sent them out to our authors as a provocation, not expecting to generate a uniform set of responses, but rather hoping to establish a framework within which the authors could develop multiple perspectives. While the majority of the essays collected here originated as talks for the Maryland Conference, they were revised for publication. To broaden the scope, three additional scholars were invited to contribute their perspectives, and they added important critical voices to the conversation. We believe the resulting volume benefits from the variety of perspectives.

While we regard the mission and work of academic and professional writing at community colleges to be critically important, we were charged with organizing a conference focused on theory, research, and best practices as they play out at research-intensive institutions, where the emphasis is on generating knowledge in the target areas. The authors represent a wide range of institutions, although the focus was on identifying outstanding scholars, regardless of institutional affiliation. The perspectives on academic and professional writing offered here, we believe, are instructional for writing teachers, scholars, and researchers at higher education institutions of all categories.

Organization of the Volume

We organized the chapters of this volume in five sections according to their primary emphasis within the broad topic of academic and professional writing: I) "Programmatic Perspectives at Maryland"; II) "Programmatic Perspectives at Illinois, Cornell, Howard, and Penn State"; III) "Language Perspectives"; IV) "Digital Perspectives"; and V) "Two Perspectives on Assessment." The afterword identifies themes across chapters and considers perspectives on the future of academic and professional writing in general.

The opening two chapters form the section "Programmatic Perspectives at Maryland." They appropriately attend to the history of the University of Maryland's two writing programs, academic and professional, and are authored primarily by those who were instrumental in developing them. In the headnote for their chapter "Innovation and Restoration: A History of Introductory Academic Writing at the University of Maryland," Robert Coogan, Jane Donawerth, and Molly J. Scanlon observe that theirs is "just one small narrative, but one that stands to enrich our disciplinary understanding of writing programs." These authors provide a history of the academic writing program at Maryland, where the idea for this collection took root. This chapter is followed by a narrative tracing the development of land-grant colleges, like the University of Maryland, and the development of rhetorically based writing programs, like the "Maryland model." Jeanne Fahnestock's essay, "The Once and Future

Discipline: The Rhetorical Core of the Academic and Professional Writing Programs," demonstrates how rhetorical principles increasingly influenced the shape of writing programs and professional societies both in the United States and around the world, as built on the rhetorical framework of stasis theory.

Section II, "Programmatic Perspectives at Illinois, Cornell, Howard, and Penn State," offers reflective histories of successes and challenges in establishing and making critical changes to the writing programs at four major institutions. The authors recount past and current programmatic initiatives developed to improve the quality of students' writing, the material conditions and preparation of writing instructors, the pedagogical and theoretical underpinnings of the writing curriculum, and the accommodations made to changing student populations. Kelly Ritter, in "Undergraduate Rhetoric at UIUC: Revising a Curriculum, Rethinking a Program," describes four challenges faced at the University of Illinois that are familiar to many large state universities: an increase in adjunct faculty, inadequate training in the teaching of writing, dual-credit programs, and a growth in the numbers of international students and multilingual learners. Teresa M. Redd recounts her experience expanding the Writing across the Curriculum (WAC) program at Howard University in which she encouraged faculty members of Howard's various schools to assume a central role in promoting good writing among the students enrolled in their courses. In her essay, "Breaking Out of the Box: Expanding the WAC Program at Howard University," Redd traces the development of the WAC program at a historically black university from its tentative beginnings in 1991 to its current state, including an up-close look at program assessment in one school. In Paul Sawyer's essay, "'Who Are the Specialists?' The Dependent Independent Writing Program at Cornell," we learn about the interdependent scene of writing program development at Cornell's Knight Institute for Writing in the Disciplines. In the closing section of his chapter, Sawyer resituates his program within the larger context of "the corporatization of the American university." Attending primarily to the powers of "feminism and rhetoric at the site of writing program administration," Cheryl Glenn traces the history of the writing program at the Pennsylvania State University in her essay, "At the Intersection of Feminism, Rhetoric, and Writing Program Administration," and concludes with reflections on the advantages and disadvantages of such a site.

The chapters in section III, "Language Perspectives," identify certain features of language that enable academic and professional communication. James Paul Gee emphasizes the interdependent relationship between system and situation, and he points to the manner in which language grows out of embodied experience in "Language as System and Situation: Writing the World." Suresh Canagarajah, in "Reconfiguring Learner Identities in Writing Pedagogy," argues for a reconsideration of the ways in which we categorize language learners in academic and professional writing courses, a shift that could influence outcome assessments as well. Returning to the challenges associated with writing program administration, Staci Perryman-Clark writes about the theoretical and practical matters associated with language rights for composition students. In her chapter, "A United Front: A Writing Program Administrator's Institutional Investment in Language Rights for Composition Students," Perryman-Clark considers these critical issues in the context of CCCC's position statement Students' Right to Their Own Language and describes strategies she initiated at Western Michigan University that may speak to writing program administrators at other institutions.

The chapters in section IV, "Digital Perspectives," attend closely to the ways in which digital practices in both academic and professional writing have shaped how writers compose and respond to texts. Charles Bazerman begins the title of his chapter with a question, "What Do Humans Do Best? Developing Communicative Humans in the Changing Socio-Cyborgian Landscape," as a way of exploring what technological affordances will enable current and future academic writers to pursue as activities that are distinctively or uniquely human. James A. Herrick identifies in the classical rhetorical tradition the roots of contemporary writing practices among the digital natives who are our students. Herrick's chapter, "From *Topoi* to Tweets: What Ancient Rhetoricians Can Teach Digital Natives," reminds us of the enduring practices of effective communication. Douglas Eyman's chapter, "Text, Design, Code: Digital Rhetoric in Academic and Professional Writing," adds to the conversation on the impact of digital rhetoric by highlighting writing's inherently visual and digital nature. In this section's final chapter, Anne Frances Wysocki's "A Beauty for Informing Digital Bodies" considers how digitality enables an approach to teaching writing that incorporates all the senses. Wysocki also addresses future writing practices in our digital age.

The last section, section V, features two chapters that (re)turn to the subject of assessment, a topic that fed our early interest in the academic and professional writing conference and led to this collection. Kathleen Blake Yancey, in "It's Tagmemics *and* the Sex Pistols: Current Issues in Individual and Programmatic Writing Assessment," promotes a view of writing assessment that takes into account not only measures of individual performance but also the inherently social aspects of writing. Yancey discusses the extent to which "writing assessment has taken its own social turn." Doug Hesse recalls his encounter with a leading proponent of the Common Core Standards in his chapter, "What If the Common Core Standards Actually Work? For Writing, Partly Cloudy with a Chance of Rain." Hesse recalls the early history of this assessment initiative and considers both favorable and unfavorable effects it could have on college writing instruction.

In the afterword, "The No-So-Simple Truth of English B," Jessica Enoch and Scott Wible place the foregoing chapters in conversation with one another, and in conversation with a famous Langston Hughes's poem, to consider what possibilities they suggest for how the field should move forward in response. We invite you to join the conversation.

Works Cited

ACT. *ACT National Curriculum Survey.* ACT, 2016, www.act.org/content/dam/act/unsecured/documents/NCS_Report_Web.pdf.

ACT. *The Condition of College and Career Readiness, 2015: National.* ACT, 2015, www.act.org/content/dam/act/unsecured/documents/CCCR15-National ReadinessRpt.pdf.

Applebee, Arthur N. *Tradition and Reform in the Teaching of English: A History.* National Council of Teachers of English, 1974.

Bartholomae, David. "Composition, 1900–2000." *PMLA*, vol. 115, no. 7, 2000, pp. 1950–54.

Berlin, James A. *Writing Instruction in Nineteenth-Century American Colleges.* Conference on College Composition and Communication / National Council of Teachers of English, 1984.

Brereton, John C., editor. *The Origin of Composition Studies in the American College, 1875–1925: A Documentary History.* U of Pittsburgh P, 1995.

Common Core State Standards Initiative. www.corestandards.org. Accessed 7 Sept. 2017.

Kitzhaber, Albert R. *Rhetoric in American Colleges, 1850–1900.* Southern Methodist UP, 1990.

Murphy, James J. *A Short History of Writing Instruction: From Ancient Greece to Contemporary America.* 3rd ed., New York: Routledge, 2012.

Murphy, Tim. "Tragedy of the Common Core: What's Pitted Glenn Beck and Teachers' Unions against Jeb Bush, Bill Gates, and the Obama Administration? The Mammoth Battle over America's Last Bipartisan Reform." *Mother Jones*, Sept.–Oct. 2014, pp. 37–68.

National Assessment of Educational Progress. U.S. Department of Education, Institute of Education Sciences, National Center for Education Statistics, www2.ed.gov/programs/naep/index.html. Accessed 1 July 2015.

Rich, Motoko. "Grading Common Core: No Teachers Required." *The New York Times*, 23 June 2015, p. A1.

Schneider, Mercedes K. *Common Core Dilemma: Who Owns Our Schools?* Teachers College P, 2015.

Strauss, Valerie. "Common Core Isn't Preparing Students Very Well for College or Career, New Report Says." *The Washington Post, Answer Sheet* blog, 9 June 2016.

Varnum, Robin. "The History of Composition: Reclaiming Our Lost Generations." *Journal of Advanced Composition*, vol. 12, no. 1, 1992, pp. 39–55.

Wozniak, John Michael. *English Composition in Eastern Colleges, 1850–1940.* UP of America, 1978.

Section I

Programmatic Perspectives at Maryland

1

Innovation and Restoration:
A History of Introductory Academic
Writing at the University of Maryland

Robert Coogan, Jane Donawerth, and Molly J. Scanlon

By reflecting on the history of the composition program at the University of Maryland to provide insights for the scholarly community on its origins and evolution of program administration, pedagogical perspectives, and academic politics, Coogan, Donawerth, and Scanlon suggest that one way to view change in writing programs is through an examination of the interaction of innovation and restoration in the field and in the program as influenced by open enrollment, national trends toward a rhetorical focus, and changes resulting from administrator and teaching assistant turnover. Much like other contributions in this collection, the work they share here is just one small narrative, but one that stands to enrich our disciplinary understanding of writing programs of the past, present, and future.

Until 1972, English 1 and 2, the University of Maryland's introductory writing courses required of all students who did not test out through SAT scores, centered on composition and American literature, current-traditional grammar, modes of discourse, and style.[1] At Maryland this approach had been fairly unchanged in focus for a dozen years. In this respect, Maryland followed the national trend, a trend that argued that "great literature" provided content for students' papers and made them better citizens (Berlin 107–09).[2] During the 1970s, caught up in the development of composition studies as a field and new views on pedagogy, the teaching of introductory writing at Maryland was transformed.[3] In response to a 1973 directive by the campus, powered by dissatisfaction with the preparation of the broader student population entering Maryland and by the hiring of Robert Coogan as the new director of then Freshman English, the transformation of the program began by refocusing on rhetoric, not literature, as the subject of the course and by enhancing the training of teaching assistants (TAs), who did almost all of the teaching of introductory writing. During the 1970s and since, Maryland has been sometimes a mirror, sometimes a leader in advances in the study of college writing, developing a robust hybrid curriculum centered on rhetoric but drawing on many schools of composition pedagogy.

The revised course of one semester was renamed English 101: Introduction to Writing, initially described as "the study and application of rhetorical principles in expository prose; frequent themes" (University of Maryland Course Catalogs [1972] 137). The 1950s and 1960s saw a revival of interest in rhetoric, and this revival transformed the teaching of composition, as suggested by the publication in 1965 of Edward P. J. Corbett's textbook: *Classical Rhetoric for the Modern Student.* A friend of Professor Coogan's, Corbett, as well as Coogan, had turned to classical rhetoric as a model for teaching writing. The Maryland program, though, was a more accessible approach than Corbett's highly technical rendition of classical rhetoric. Maryland's curriculum centered on Aristotle's three appeals (ethos, logos, and pathos) and, for generating a thesis and supporting arguments, an inventional system based on the commonplaces[4] (as in the modernized version in the textbook *Twenty Questions for the Writer* by Jacqueline Berke). The University of Maryland course also covered grammar but in the form of style and rhetorical effectiveness, as well as modern linguistics. Such an approach, Coogan explained in an interview, is "not subject-centered; it is method-centered," and it allows students "considerable leeway in choosing the subject matter," while also requiring them to consider "the position you want to take," "the lines of the topics," and "how to appeal to the various audiences."

The training of TAs also mirrored an increased emphasis on pedagogy, an emphasis unique to writing programs at the time, since professors generally assumed that one knew how to teach literature from having made it through twenty years of instruction. But entering graduate students were often completely unaware of the long tradition of teaching persuasive writing, and most had never considered the *how* of effective teaching. During the summer, entering TAs received an orientation schedule, the name of their assigned "master teacher," the textbook they would use, and the department's grading standards along with a sample set of papers to be graded. TA training at Maryland included an orientation of several days before the semester began in the fall and a semester-long course, Approaches to College Composition. Orientation, with presentations by master teachers, covered prewriting, the nature of the paragraph, and the construction of topic sentences. The seminar, initially co-taught by three professors, offered an introduction to standards of grammatical usage and aspects of modern linguistics' understanding of language, a close reading of Aristotle's *Rhetoric* and an overview of the development of classical rhetoric and its influence, a consideration of I. A. Richards's modern update of rhetoric (with sessions on the role of definition and modern theories of metaphor), and an introduction to logic. Master teachers, assigned four novices each, supplemented the seminar with weekly meetings to discuss lesson plans, ways of presenting new material, the careful grading of papers, and problems with students. Because of the master teachers' enthusiasm, a close collegial relationship existed among the teachers of introductory writing (Coogan, "And Gladly").

As the program adjusted to better train novice TAs, the population of students entering college was diversifying due to open enrollment. By the 1970s, the growing number of students enrolled in English 101, many of them not well prepared in writing skills, constituted a challenge to the teaching of introductory writing (see Faigley, especially 27–29; and Bizzell). For example, of the 1,327 first-year students entering in 1973, 42.5 percent of black students and 15 percent of white students scored below 401 on the verbal portion of the SAT.[5] At the time, the university required English 101 students to pass a usage test in order to complete the course; unprepared students were not able to pass, resulting in a 30 percent failure rate. The Introductory Writing Program addressed this problem in several ways. First, Coogan designed a modular program—English 104, 105, and 106—that substituted for English 101 for students entering with scores below 401 or for students identified by teachers of English 101 as possessing inadequate grammatical skills. Learning first about the sentence, then the paragraph, then the essay, students had to pass the final in one module before taking the next, but they did not have to pay for extra credits; such a system was not remedial, since it did not require students to take an extra course—and most students finished the series by the middle of second semester.[6]

Also, because of the success of the civil rights movement in the 1960s, the racial balance of the state of Maryland, and the efforts at the University of Maryland to attract a greater number of students from underrepresented groups, there were increased numbers of African American students attending a university that had been one of the last major state universities to desegregate. The Introductory Writing Program needed to make these students, often first-generation college students, feel welcome. In 1974 Dr. Orlando Taylor of the Center for Applied Linguistics provided graduate teachers a workshop on Ebonics, the study of Black English, so that teachers had a perspective from which to grade and to discuss papers written in nonstandard English.

The Introductory Writing Program, in affiliation with the Commission on Minority Student Education, assigned teachers to Independent Educational Development, a program to improve the instruction of black students, and hosted a series of lectures featuring speakers such as Wallace Terry, a black journalist who covered the Vietnam War. Thus when Coogan stepped down from his position as director of the reformed Introductory Writing Program, he left a legacy that put the University of Maryland ahead of most other U.S. universities: a significant investment in educating TAs in teaching composition and rhetoric, a curriculum based on finding the best means of persuasion, and a program that acknowledged and welcomed students from diverse cultural and linguistic backgrounds.

In 1974 the Introductory Writing Program also established a writing center where students seeking individual help were tutored; run by three graduate students and ten undergraduates with special training, the center served a university-wide clientele eight hours a day, five days a week. In 1979 Dr. Leigh Ryan became coordinator of the writing center, and from 1982 to the present, director. By 2015 the writing center had expanded to seventy tutors who worked in a customized space in the Department of English on the ground floor of Tawes Hall; today peer tutors are trained in a semester-long internship course that now includes both face-to-face and online strategies (see Ryan and Zimmerelli). Tutors (peer, graduate student, and retired volunteers) conduct face-to-face and online tutoring and consult by telephone for the grammar hotline, and tutors, graduate student assistant directors, and the director conduct workshops across campus, regionally at many high schools, and even internationally—the director has consulted in Ireland, Germany, South Africa, and Japan, among others.

Eugene Hammond followed Bob Coogan as director of the Introductory Writing Program. His 1985 textbook, *Informative Writing*, and his 1983 book, *Teaching Writing*, give a clear picture of the additions to the English 101 curriculum from developments in expressivist rhetoric and cognitive studies in composition during the late 1970s and early 1980s. In an interview, Professor Hammond recalled, "It was so exciting because every six months or every three months there was a new and really interesting book." To the central focus on classical rhetoric (especially ethos, logos, and pathos and the logical fallacies), Hammond added emphasis on writing as a process and revision, and an expressivist concern for telling details and individual voice. The program reduced the number of graded papers from eight to six, mixing in some short ungraded writing. The program also dropped the themes in the modes of discourse, reorganized the basic writing course into a semester-long section of

English 101,[7] and changed the terminology of *master teacher* to *mentor teacher*. In English 611: Approaches to College Composition and in new-teacher orientation, composition pedagogy was emphasized, and new TAs discussed in-class writing, journals, group work, and peer critiques.[8] Hammond took over responsibility for the whole of English 611 and insisted on teaching introductory writing once a year: "I think it's foolish," Hammond professed in an interview, "for anybody to teach people to teach something that you don't do." It became the responsibility of the director to visit and discuss with TAs a class they taught during their first or second year in the program, and mentor teachers frequently observed the classes of the TAs they supervised. The development of the mentor teacher program not only encouraged accountability but also reinforced reflexive pedagogical practices.[9]

With the success of the Introductory Writing Program, but responding to public anxiety about college students' writing skills, the University of Maryland established the Professional Writing Program for students at the junior level in 1980. One of the original models for upper-division writing programs nationally, the Professional Writing Program began as a writing-across-the-curriculum program but soon rebounded into the English department, under the directorship of Michael Marcuse. With a basic focus on rhetoric, the program at first consisted of two courses: English 391 (professional writing) and 393 (technical writing). The program has expanded with increasingly specialized courses and, by 2015, was offering over thirteen courses, including business, health, legal, science, environmental, and humanities writing. Many students take more than one course.[10]

From 1984 to 1992, the curriculum of the Introductory Writing Program was added to, rather than redesigned. Under the influence of the Departments of Women's Studies and African American Studies at the University of Maryland, there was greater attention paid to curricular transformation and inclusiveness in the classroom. As director of Introductory Writing from 1984 to 1988, Jane Donawerth emphasized classroom climate and organized mentor teachers for a revision of course materials for ESL (English as a second language) sections of English 101. Also added during this period were training in adult cognitive development for TAs (who could then better understand first-year students' obsession with grades) and a computer lab for English 101 teachers. In an interview, Donawerth stated her goals for Introductory Writing: "it's important that students know that they have choices . . . that they have a repertoire rather than only one method . . . so that they can see that there aren't regulations in writing, that there are only rules of thumb, that there are rhetorical guidelines, that they have rhetorical choices." Succeeding

Donawerth in 1988 was Dr. Leigh Ryan, director of the writing center. During her tenure, Ryan added "conferencing theory, about talking with students about their writing," as she explained in an interview. Subsequently, Professor John Schilb, director from 1990 to 1992, did away with the outdated grammar proficiency examination, which he described in an interview as "an insufficient gauge of students' actual writing ability." Throughout the 1970s, 1980s, and up to the 1990s, directors and mentor teachers supervised seventeen to thirty-four new TAs each year, and directors taught English 101 and English 611 every year, reviewed teaching evaluations for all new TAs after their first year, visited a class taught by TAs during their second year, and eventually wrote a "teaching letter" for the job files of TAs when they finished their PhD or MFA.

During the 1990s, the department developed interests in rhetoric and composition in ways beyond the writing programs. From the late 1980s until the late 1990s, when the department did away with concentrations, the concentration in language, writing, and rhetoric was the second-largest set of majors, after English and American literature. In addition, faculty in communications and English joined to create a rhetoric minor that was approved in 2005 and continues today, graduating from thirty to fifty students every year.[11]

By the time the Introductory Writing Program reached the 1990s, the curriculum centered on rhetoric, but rhetoric hybridized with new developments in the field of composition and rhetoric, so that English 101 offered teachable ways of generating arguments, organizing materials, and appealing to an audience; an approach to the process of writing influenced by cognitive psychology; and a view of problem solving and inventing arguments as creative aspects of the composing process (see Berlin 157–61). The classical rhetorical emphasis on persuasion was modified by a modern understanding of truth as a social negotiation, a dialectical interaction with the audience (Berlin 168–70).

But at the same time, the focus on rhetoric had blurred, especially under the influence of the growing importance of literary theory in English studies. Part of the history of the curriculum at the University of Maryland, then, is restoration. In her brief tenure as director of writing programs and acting director of the Introductory Writing Program from 1992 to 1994, Jeanne Fahnestock restored the focus on classical rhetoric by requiring all teachers to teach from a syllabus template centered on rhetoric, and by publishing a handbook of assignments, explanations, and exercises for students in 101. In an interview, Professor Fahnestock explained, "Because TAs in English know relatively little about language and rhetoric, they are always tempted to go back to reading [literature] and putting texts in first place in the course, rather than spending the time on writing tactics and practice. Without readings, they feel the course has no 'content.'" Fahnestock's syllabus introduced to English 101 the teaching of stases for generating arguments,[12] and it set up a required linked series of rhetorical essays for students: an encomium, a definition, a rhetorical analysis, an argument on both sides of the issue, and, finally, an extensive researched proposal paper.

The directors who followed in the 1990s maintained the program's historical focus on rhetoric. Professor Linda Coleman, whose specialty was linguistics, brought attention to contemporary understandings of grammar back to TA training and also formalized the 101 handbook through custom publication of *Introduction to Academic Writing*. She further organized mentor teachers to edit *Perspectives*, a collection of published writings and student papers that was updated each year. Dr. Linda Macri kept the focus on rhetoric and stasis theory, as well, but changed the first paper to a personal experience assignment and dropped the number of graded assignments to four, treating some of the assignments as ungraded in-class group work. During the 1990s, the class size had crept up to twenty-five under budgetary pressure, and Macri campaigned to bring it back down to twenty-two again.

With a new chair who was very supportive of the writing programs, a departmental writing committee was established in 2008, and Professor Shirley Logan was appointed to the new position of director of writing programs, almost immediately also setting up a University of Maryland Campus Writing Board. The English department writing committee's first responsibility was a review of the curriculum of both the Introductory Writing Program and the Professional Writing Program. Under the auspices of the committee and its review, the introductory program changed its name to the Academic Writing Program, upped the number of graded assignments to five, dropped one of the by-then two personal experience papers for a beginning summary of a research article (as a result of an informal campus survey about expectations for what students should know after 101), and began to encourage teachers to experiment with assignments using visual presentation software and other digital modes of composition.

Under Professor Jessica Enoch, director of the Academic Writing Program from 2013 to the present, and with the encouragement of the writing committee, the program restored the number of assignments to six papers, adding a web page assignment,[13] in which students set out different parts of an issue with an eye to a public audience, and a final revision and reflection paper that aided students in understanding how they might transfer what they had learned to new writing projects. The course as a whole emphasizes writing as inquiry but retains the deep investment in rhetoric of the Academic Writing Program and the hybridity that encourages absorption of the current conversation on composition pedagogy into the program. From Coogan's time to the present, coordinators and mentor teachers have been important collaborators in the 101 curriculum, putting together the first basic writing handbook, choosing textbooks, organizing files of assignments and exercises, aiding in revising syllabus templates, editing *Perspectives*, and beginning an annual online collection of select student essays. One coordinator developed a sophomore-level course in service learning to add to departmental offerings on writing, which span every level. Importantly, the Academic Writing Program office under Enoch became a site for research on transference and adaptation from one genre to another, and the director and coordinators of the program presented their results at the March 2015 Conference on College Composition and Communication.[14] Teachers in the Academic Writing Program have moved closer to the ideal set forth by Robert Connors: "scholars who embrace teaching and service as indispensable parts of the world of their research, and put scholarly research in the service of action" (20).

The Academic Writing Program continues to reflect or even to lead in advances in the field of college writing. The curricular mainstay of rhetorical approaches to writing has been preserved, and as the student population has diversified, the program has expanded and changed to include new approaches to teaching college writing.

Appendix: Directors of the University of Maryland Academic Writing Program

Jack Barnes	1958–1968
Harold Herman (interim)	1968–1971[15]
Joanna Schmeissner (interim)	1971–1972[16]
Robert Coogan	1972–1978
Eugene Hammond	1978–1984
Jane Donawerth	1984–1988
Leigh Ryan	1988–1990
John Schilb	1990–1992
Jeanne Fahnestock	1992–1994
Linda Coleman	1994–2004
Linda Macri	2004–2013
Jessica Enoch	2013–present

Notes

1. Wilcox suggests that many institutions had a similar curriculum up into the 1970s: "readings in non-literary expository prose, discussion of verbal patterns and rhetorical strategies, and a sequence of writing assignments" (690). See Scanlon for an appreciative account of Jack Barnes's years as director of this program, 1958–70.

2. Berlin also suggests that in the Cold War climate of witch hunts, literature and composition provided a "safe" subject, not politically dangerous, for college writing courses and rhetoric (107). At the time, "great literature" was taught as if free of ideology.

3. See Connors: "The 1970s, rather than the 1960s, were the founding decade of the disciplinarity of composition studies" (8).

4. The topoi, topica, loci, topics, or commonplaces are a set of categories invented to generate arguments by asking questions: for example, what sort of argument based on definition would support my claim? See Aristotle (142–59; book 2, ch. 23), Cicero (vol. 2, 317–19; book 2, sec. 39), and Quintilian (vol. 2, 213–53; book 5, ch. 10).

5. Email from Dr. Robert Coogan, 28 February 2015.

6. On the history of basic writing instruction, see *A Sourcebook for Basic Writing Teachers*, edited by Theresa Enos, especially essays by Lynn Quitman Troyka, David Bartholomae, Mike Rose, and Andrea Lunsford.

7. On the rethinking of basic writing to emphasize overall cognitive skills rather than modular units on sentences and paragraphs as best practices, see Lunsford (254).

8. Such a pedagogy also fits with expressivist composition practice of the time; expressivists believed that "writing can be learned but not taught," and they developed a set of teaching strategies to nurture beginning writers and to elicit a writer's authentic voice (see Berlin 145–55; Murray).

9. Reflexivity is the habitual incorporation of reflection into one's practice. Formerly an extension of feminist theory and research methodology, reflexivity has been increasingly incorporated into composition pedagogy. See Donna Qualley's *Turns of Thought: Teaching Composition as Reflexive Inquiry*.

10. On the Professional Writing Program, see the essay by Jeanne Fahnestock in this volume.

11. Email from Professor Vessela Valiavitcharska, recent University of Maryland advisor for the rhetoric minor, 6 April 2015.

12. On the modernized form of the stases—definition, cause, value, proposal—taught at Maryland, see Fahnestock and Secor.

13. On the institutional changes needed to sustain a digital composition program—not only technologically enhanced classrooms but also revision of composition learning outcomes, digital information in the TA teaching guide, and digital training in the composition pedagogy course—see Adsanatham and coauthors. On the pedagogical goals of a digital assignment—developing multimodal strategies for revision, adapting rhetorical strategies to new media, and pursuing new critical literacies to evaluate the results—see Palmeri (149–59).

14. On prompting students through reflective writing to categorize their writing into their own idiosyncratic metagenres, as a means of enabling them to transfer genre knowledge across contexts, see Lindenman's study.

15. Interim Director Harold Herman (1968–71) succeeded Jack Barnes. Herman was a faculty member in literature, who retired as professor emeritus in 1994.

16. Interim Director Joanna Schmeissner (1971–72) succeeded Harold Herman. Schmeissner was an instructor for communication courses and went on to hold several administrative support positions in the university, including in the Office of Academic Affairs and the Office of University Marketing and Communications. She has since retired.

Works Cited

Adsanatham, Chanon, et al. "Going Multimodal: Programmatic, Curricular, and Classroom Change." *Multimodal Literacies and Emerging Genres*, edited by Tracey Bowen and Carl Whithaus, U of Pittsburgh P, 2013, pp. 282–3120

Aristotle. *Rhetoric*. Translated by W. Rhys Roberts, Modern Library of Random House, 1954.

Bartholomae, David. "Teaching Basic Writing: An Alternative to Basic Skills." *A Sourcebook for Basic Writing Teachers*, edited by Theresa Enos, Random House, pp. 84–103.

Berlin, James. *Rhetoric and Reality: Writing Instruction in American Colleges, 1900–1985*. Southern Illinois UP, 1987.

Bizzell, Patricia. "What Happens When Basic Writers Come to College." *College Composition and Communication*, vol. 37, no. 3, 1986, pp. 294–301.

Cicero. *De oratore*. Translated by E. W. Sutton and H. Rackham, Harvard UP, 1967. 2 vols.

Coleman, Linda. Interview by Molly Scanlon. 12 May 2008.

Connors, Robert J. "Composition History and Disciplinarity." *History, Reflection, and Narrative: The Professionalization of Composition, 1963–1983*, edited by Mary Rosner et al., Ablex Publishing, 1999, pp. 3–21.

Coogan, Robert. "And Gladly Would They Learn, and Gladly Teach." *University of Maryland Graduate School Chronicle*, Nov. 1977, pp. 5–7.

———. Interview by Molly Scanlon. 2 May 2008.

Corbett, Edward P. J. *Classical Rhetoric for the Modern Student*. Oxford UP, 1965.

Donawerth, Jane. Interview by Molly Scanlon. 13 May 2008.

Enoch, Jessica. Interview by Jane Donawerth. 27 Mar. 2015.

Enos, Theresa, editor. *A Sourcebook for Basic Writing Teachers*. Random House, 1987.

Fahnestock, Jeanne. Interview via email by Molly Scanlon. 12 May 2008.

Fahnestock, Jeanne, and Marie Secor. *A Rhetoric of Argument*. Random House, 1982.

Faigley, Lester. "Veterans' Stories on the Porch." *History, Reflection, and Narrative: The Professionalization of Composition, 1963–1983*, edited by Mary Rosner et al., Ablex Publishing, 1999, pp. 23–37.

Hammond, Eugene R. *Informative Writing*. McGraw-Hill, 1985.

———. Interview by Molly Scanlon. 20 Apr. 2008.

———. *Teaching Writing*. McGraw-Hill, 1983.

Lindenman, Heather. "Inventing Metagenres: How Four College Seniors Connect Writing across Domains." *Composition Forum*, vol. 31, Spring 2015. Web.

Lunsford, Andrea. "Politics and Practices in Basic Writing." *A Sourcebook for Basic Writing Teachers*, edited by Theresa Enos, Random House, 1987, pp. 246–58.

Macri, Linda. Interview by Molly Scanlon. 22 Apr. 2008.

Murray, Donald. "Internal Revision: A Process of Discovery." *Research on Composing*, edited by Charles R. Cooper and Lee Odell, National Council of Teachers of English, 1978, pp. 85–103.

Palmeri, Jason. *Remixing Composition: A History of Multimodal Writing Pedagogy*. Carbondale: Southern Illinois UP, 2012.

Qualley, Donna. *Turns of Thought: Teaching Composition as Reflexive Inquiry*. Boynton/Cook, 1997.

Quintilian. *Institutio oratoria*. Translated by H. E. Butler, Harvard UP, 1966. 4 vols.

Richards, I. A. *The Philosophy of Rhetoric*. Oxford UP, 1965.

Rose, Mike. "Remedial Writing Courses: A Critique and a Proposal." *A Sourcebook for Basic Writing Teachers*, edited by Theresa Enos, Random House, 1987, pp. 104–24.

Ryan, Leigh. Interview by Molly Scanlon. 1 May 2008.

Ryan, Leigh, and Lisa Zimmerelli. *The Bedford Guide for Writing Tutors*. 6th ed., Bedford/St. Martin's, 2015.

Scanlon, Molly. *A History of the Introductory Writing Program at the University of Maryland (1958–1978)*. U of Maryland, MA thesis, 2008.

Schilb, John. Interview via email by Molly Scanlon. 20 May 2008.

Troyka, Lynn Quitman. "Perspectives on Legacies and Literacy in the 1980s." *A Sourcebook for Basic Writing Teachers*, edited by Theresa Enos, Random House, 1987, pp. 16–26.

University of Maryland Course Catalogs. U of Maryland, 1958–1972, University Libraries Special Collections, www.lib.umd.edu/univarchives/catalogs. Online archive.

Wilcox, Thomas W. "The Varieties of Freshman English." *College English* 33, no. 6, 1972, pp. 686–701.

2

The Once and Future Discipline: The Rhetorical Core of the Academic and Professional Writing Programs

Jeanne Fahnestock

Courses in rhetoric and composition have been a consistent feature of the University of Maryland's curriculum from the 1860s, through the formation of multiple colleges in the 1920s, and enrollment expansion after World War II. During the 1960s and 1970s, writing courses lost ground in many universities, but Maryland's first-year writing course was reformed on the principles of classical rhetoric, and in the early 1980s, a new upper-division writing program was added. This "Maryland model," as James Kinneavy called it, is staffed by professional writing teachers rather than faculty in other disciplines, and writing courses are grounded in rhetorical principles. The current growth of academic and professional writing programs like Maryland's demonstrates that rhetorical education will be even stronger in the future.

In 2013 the prestigious journal *Science* published a special issue on STEM (science, technology, engineering, and mathematics) education that included an article promoting curricular goals in science for K–12 students in the United States (Stage et al.). The article featured a large three-circle Venn diagram depicting in two of its circles selections from the Common Core State Standards in mathematics and English language arts, and in its third circle proposed Next Generation Science Standards. The Common Core Standards in math and literacy have been the focus of wide-ranging criticism in many forums and probing questions about their implementation (see Hesse's chapter in this collection). But worth noting in this particular *Science* article is the prominent visual argument created by the authors' use of a three-circle Venn diagram, a visual form first proposed in 1880 by John Venn for depicting and testing inferences, now a familiar template to readers of *Science* (Venn).

Though represented visually as one of three equal circles in the diagram, the science standards did not really have the status, the wide acceptance, of the other two sets at the time (Mervis). The mathematics and language arts standards were adopted by forty-five states as of 2013; the science standards were then under development by the National Research Council and twenty-six state partners. But equal status is what the authors actually want, and the equal circles "argue" for that status. In further pursuit of their visual argument, the authors placed science and math in the top two circles, a yoking appropriate to their audience of *Science* readers. And they then adapted to the narrow segments created by the intersecting circles with abbreviated phrasings of some of the standards. Since only a few short phrases fit legibly in the constrained spaces, the authors selected and paraphrased those most useful to their case. In this way, the chosen form of the three-circle Venn diagram literally shaped their argument.

One of the segments the authors had to fill is the center space where all three circles overlap. Occupying this core of the core, as common to science, math, and the language arts, are goals that will sound familiar to writing teachers: "Read, write, and speak grounded in evidence. . . . Construct viable arguments and critique reasoning of others. . . . Engage in argument from evidence" (276). This emphasis on viable argument (that is, on argument that will actually convince someone), and on evidence and on engaging others, could have been summed up under the term *rhetoric*—for 2,500 years the art of argument based on the available means (norms of reasoning and evidence) in each particular case. But the *Science* authors avoid the *R* word, which rarely appears with a positive connotation in the pages of *Science*. Indeed, it is possible to do mini corpus studies in online archives these days. Such a search reveals that, between 1883 and 2010, *rhetoric* and its derivative forms appear almost 1,200 times in *Science*, and a sampling of these shows overwhelmingly the predictable negative sense of *mere rhetoric*, of rhetoric on the wrong side of the appearance/reality dichotomy. So avoiding the word *rhetoric*, deliberately or intuitively, was a wise rhetorical "choice" on the authors' part. Teaching this kind of audience accommodation through a wise selection of terms, as well as the means of multimodal persuasion on display in the authors' Venn diagram—and someday, perhaps, even changing the connotation of *rhetoric* in the pages of *Science*—these are the goals of academic and professional writing programs like those at the University of Maryland and elsewhere, the subject of this collection of essays.

Like the previous chapter, mine will focus on the University of Maryland's writing programs, and it will first be more celebratory, or epideictic, followed by a few deliberative observations and even classroom suggestions. I have the great good fortune to be an insider formerly associated with both of Maryland's programs. So I want to speak first about their origins and unique institutional circumstances, since every writing program takes local features from its setting, as well as its place in larger national and historical trends. In my overview, I will revisit some history—history from thirty to forty years ago, one hundred and sixty years ago, and even five hundred years ago—and along the way, as in the opening, I will illustrate the usefulness of the rhetorical principles that ground the writing programs at the University of Maryland.

Historical Moments in Maryland's Academic and Professional Writing Programs

Maryland's Academic Writing Program, with its freshman composition course, English 101, is the older program, its identity tied to the nature of the university's overall mission. And it is worth remembering that Maryland began as Maryland Agricultural College in 1856 and was later endowed by the Morrill Act of 1862, which established a new form of publicly funded higher education in the United States. The Morrill Act granted federal land to states for

> the endowment, support, and maintenance of at least one college where the leading object shall be, without excluding other scientific and classical studies, and including military tactics, to teach such branches of learning as are related to agriculture and mechanic arts, in such manner as the legislatures of the State may respectively prescribe, in order *to promote the liberal and practical education of the industrial classes in the several pursuits and professions in life.* (Morrill Act, sec. 4; italics added)

The Morrill Act was the result of a persuasive campaign in the 1850s for new forms of higher education in the United States, including, as one high point in this campaign, Jonathan Baldwin Turner's argument for "Industrial Universities for the People" on behalf of the Industrial League of Illinois (Turner). Turner, a former professor of rhetoric at the Congregationalist Illinois College, called for new schools that would first create new knowledge for mechanics, farmers, merchants, and artisans and then take it to the people (Brown). On the national level and using different appeals for a different audience, Justin Morrill of Vermont argued the same case successfully before Congress, culminating in an 1858 speech that also circulated as a pamphlet (Morrill). Unlike Turner, who argued for broad industrial education, Morrill focused on agriculture, and we can analyze how he makes his case by using a rhetorical reading frame.

Asking first, "What are the facts?" (4), Morrill answers by assembling evidence for the declining productivity of U.S. farmland from 1840 to 1850. (The printed version of his speech features a data table [4]—a visual format for presenting evidence about two hundred years old at this point.) Establishing a decline in agricultural productivity in the older states was not an easy argument to make at the time since commodity prices were relatively stable given the new lands coming under cultivation. But Morrill, in an early appeal for sustainability, was after the broad picture of Americans using up the land and moving west. He traces the cause of declining fertility in both northern and southern states to a lack of scientific principles of land use (6–7), including what he calls in the idiom of his day a "philosophy of manures" (8), and this lack of knowledge, in turn, to a lack of schools to discover and teach the needed principles (9). Morrill assesses the negative moral and practical consequences of this agricultural ignorance (7), and he proposes the creation of agricultural colleges by grants of federal land to the states (12). The final portion of his speech is an extended argument for the legal precedents and, especially important in 1858, the constitutionality of his proposal (12–15). The outline of this speech—covering the facts of agricultural decline, its causes in ignorant farming practices and lack of knowledge and education, its evaluation as morally wrong and practically damaging, and its remedy in a proposal for new schools—follows overall the rhetorical *stases*, the list of higher-order questions concerning fact, cause, value, and action that can be addressed either singly or sequentially in arguments (Fahnestock and Secor). Using these rhetorical principles as a reading frame elucidates the content of Morrill's verbal argument, just as rhetorical principles explain the visual argument supporting the *Science* article on STEM education, an argument also concerned with the best possible form of technical education for U.S. citizens.

The actual arrangement or sequence of parts in Morrill's speech, which opens with a reference to constitutional issues (3), reflects its place in an ongoing debate. The style of his speech runs the gamut of registers and devices appropriate to its different appeals. It would, in fact, be possible to illustrate an entire course in rhetoric from this speech alone, since along the way Morrill uses every device in the armamentarium of the rhetorical training he received at Thetford Academy in Vermont: quoting the authority of Washington and Jefferson on the importance of agriculture to the nation (13), making envy-rousing comparisons to European practices (10), and anticipating the objections of established schools in the United States. "There would be no clashing of interests," he says. "Our present literary colleges," he says, "need have no more jealousy than a porcelain manufactury would have of an iron foundery [*sic*]" (8).

But there was a clashing of interests since the nature of this new form of higher education, the form we now take for granted, was an unknown at the time. Turner in Illinois had called for "some *practical, liberal* system of education for the industrial classes" (14; italics added). The wording of the Morrill Act also yokes the "liberal and practical" but does so in a compound phrase subject to varying interpretations depending on how separate or synonymous those two coordinated adjectives seem. And of course we have to recapture mid-nineteenth-century senses of *liberal* in this context, suggesting the traditional liberal arts, the education of a gentleman, and preparation for the learned professions.

It proved difficult at first to translate this passage from the bill into action, as the Morrill funds were used in different ways in different parts of the country. My thanks to John Brereton, historian of writing programs, who shared an unpublished talk with me tracing the implementation of the Morrill Act in East Coast schools as the relationship between the "liberal" and the "practical," or "vocationalism," worked itself out from state to state ("Liberal Arts" and *Origins*). In New England, for instance, the Morrill endowment went to Yale, Dartmouth, and Brown. Yale and Dartmouth created separate schools with different entrance requirements, set apart in various ways from their existing faculty and student body. At Yale, students in the science school could not earn regular university degrees; at Dartmouth, students of the practical arts were not allowed to sit with the regular students. Brown apparently sat on the money ("Liberal Arts"). Eventually, thanks in part to complaints from Farmers' Granges (local units of a national organization) in these states, separate schools were established—namely, the Universities of Connecticut, Massachusetts, and Rhode Island. At Maryland and Michigan State, existing attempts at agricultural schools were funded. Illinois, home of the activist Industrial League, eventually created an entirely new university, as did other midwestern states.

The history of the Morrill-funded Maryland Agricultural College, as Brereton shows through a history of its presidents, was to undo and redo the curriculum and the school's organization under the conflicting interests of its presidents, the board of trustees, the state legislature, and citizens' groups ("Liberal Arts"). One of the first presidents, Samuel Register, formed the school on the model of a private, denominational East Coast college by eliminating most of the science and agricultural courses, until he was ousted by the board of trustees. This traditional emphasis can be seen in the 1865 college catalog, which features studies of the classics in Latin and Greek.[1] Later presidents went the other way, eliminating most of the liberal arts courses in favor of agriculture and engineering, as sampled in the 1875 catalog, where the sciences are emphasized and Latin is now optional. But in both the traditional and the new science-dominated curricula, courses in rhetoric, composition, logic, elocution, declamation, and other liberal arts persist across all four years until the early twentieth century (see, for example, the science versus arts curricula for 1874). The current form of one course in the freshman year and one in the third year is on display in the 1918 catalog.

Maryland's presidents in the 1920s and 30s, Albert Woods and Raymond Pearson, both from midwestern land-grant schools, created the multicollege university recognizable today. But the tensions between the liberal and practical, mandated into coexistence by the Morrill Act, have never really gone away. And the course that sits at the center of that tug-of-war is the university-level writing course, the requirement that survived in both the early traditional and technical versions of Maryland.

English 101, which first appears under that number in the course catalog for 1918–19, was for decades a yearlong requirement, similar to such courses at other large public institutions. (We do not have a history of those years.) In the post–World War II Americanization environment (Berlin 92, 109–11), the writing requirement was filled by a two-course sequence in American literature, and it was strained, like all schools at the time, by the scale of delivery to ever-larger student populations. The GI Bill increased the enrollment at Maryland from three thousand to thirteen thousand in just two years.[2]

Fortunately for Maryland, the freshman composition program survived a time of crisis, curricular as well as political, in the late 1960s. In 1968 the Association of Departments of English attempted a nationwide survey of composition programs, and the feedback included a report titled *Nine Institutions Which Have Eliminated the Course in Freshman Writing*. The report names the University of Maryland as one, but that is actually the University of Maryland, Baltimore County (UMBC). The rationale offered by UMBC's director was that students who used traditional readers had nothing to write about—or rather nothing interesting to the teacher to write about—and would be better served by a traditional course in literature (Nelson 17). There is no mention anywhere of any rhetorical or even discourse skills that should be offered. A higher-profile case of the turn that could have been taken was the University of Wisconsin's decision to eliminate freshman composition in 1969, a decision that David Fleming has analyzed in his history of *Freshman Composition and the Long Sixties, 1957–1974*. The immediate reason at Wisconsin was discontent among teaching assistants over the perceived irrelevance of the course material against the background of Vietnam War. Wisconsin's English department voted to abolish the program even though it was a university requirement. It was restored in 1996.

But College Park, facing similar pressures at the time, chose to keep its first-year course yet reform it. We know more about Maryland's Academic Writing Program in the 1960s and 70s thanks to the work of Jane Donawerth's student Molly Scanlon, who investigated its archives and took oral histories from its directors. In the early 1970s, the dean of undergraduate studies told the English department of mounting dissatisfaction with freshman writing and a coming review. At a time of expanding enrollments, the course was failing a third of its students. The then department chair, Shirley Kenney, brought in a new director, Robert Coogan, an experienced high school teacher and scholar of Renaissance humanism, who gave the program a marked rhetorical core (Scanlon 15–16). To prepare teaching assistants, Coogan created a required graduate course, English 611, still in existence today (though now taken in the semester before the teaching assistant teaches English 101 to undergraduates). Coogan's 611 was one-third on logic or dialectic, one-third on language, and one-third on rhetoric—in short, it replicated the trivium. Bob Coogan also created a writing center staffed at first with retired volunteers—now a global model under its director Leigh Ryan. Coogan's rhetorical orientation persisted through many curricular changes in emphases and is still visible in the current program under Jess Enoch.

The Professional Writing Program also owes its origins to the dissatisfactions of the 1960s and 70s, as well as to a perception of students' writing problems in upper-division courses and, added to the mix, complaints from employers at the time about the poor preparation of graduates. So the exigence was strong at Maryland to "do something," and influential in understanding the problem at the time was the Dartmouth Study of Student Writing described by Albert Kitzhaber in the early 1960s (see Kitzhaber). The study documented a decline in undergraduate writing across four years of college as measured primarily by error assessment, and Kitzhaber presented the evidence in a table, "Rate of Errors per 1,000 Words" (109), showing once again how a form like a data table invites the creation of the kind of data that can fill it. In analyzing the results, Kitzhaber argued that no course taken in the first year could be expected to have an influence over four years without reinforcement. So more improvements to the first-year course were not the answer. In retrospect, the perceived failures of upper-division students, or of graduates new to the job, can also be explained as the predictable loss of competence that occurs when learners face more challenging tasks from new material, situations, and genres. Educational theorists call this U-shaped learning. But then, as sometimes now, writing was defined as a one-size-fits-all skill that can be assessed in a one-time test.

At Maryland, with complaints from within and without, and with the views of Kitzhaber's and similar studies in mind, the faculty senate created an all-campus advisory committee in 1973 composed of faculty from across campus (Marcuse 1.A.)—in effect, a body like the Campus Writing Board that sponsored the 2014 Maryland Conference giving rise to this collection of essays. In its final report released three years later, this faculty advisory committee took the bold step of recommending a required advanced writing course taken during the junior year "in order to reinforce and refresh writing skills at the time the student is entering upon advanced undergraduate work" (Marcuse 1.A.). The faculty senate voted for the proposed change in the requirements on September 20, 1977, and "Junior Writing" came on line for the class entering in 1980 under the directorship of Michael Marcuse—who came back after 2000 to oversee a dramatic expansion in the program's offerings.[3]

Thanks to Jim Kinneavy, we can understand the place of Maryland's new requirement in the national landscape of writing programs in the early 1980s. In an article published in 1983, he surveyed the territory covered by the then relatively new term "writing across the curriculum." Today of course that term applies to a wide variety of practices including writing-intensive courses, writing in the disciplines, and many more. But in 1983 Kinneavy was talking in general about expansion into upper-division courses, an expansion that could, he said, be "modeled on either the Michigan or the Maryland variation" (17). He called the Maryland variation the "centralized" writing department approach ("Writing" 14), which offered general courses in advanced composition and technical, scientific, or business writing, contrasted with the Michigan model, which at the time emphasized writing courses within majors taught by the faculty in a discipline. (In the early 1980s Kinneavy's home institution, the University of Texas at Austin, had courses like Maryland's, though not a complete requirement.) Although he pointed out the strengths and weaknesses of both approaches and wanted both versions instituted at Texas, one in the third and one in the fourth year, Kinneavy had much to say about some of the benefits of the centralized Maryland model. First, it created a staff of dedicated writing teachers, highlighting the importance and independence of the discipline and its classes. Second, he thought that it was extremely important for specializing students to practice addressing general readers since doing so helped to create responsible and ethical professionals and improved democratic participation (18). Third, he saw the centralized approach as the restoration of a rhetorical core across disciplines, though at the same time he also saw the need for research into the methods of argument preferred in different disciplines for both variations. Of course that kind of comparative research into methods of argument requires once again a common discipline of analysis to even come into view. That common discipline, as Kinneavy well knew, was already available in the arts of the trivium, and he wrote at length elsewhere on the revival of rhetoric as the architectonic art subsuming its partners, grammar and dialectic ("Restoring"). Maryland's Junior Writing Program, renamed the Professional Writing Program in 1989, was an attempt, at least on paper, to follow Kinneavy's recommendations. (And I will come back to some of the Kinneavy-inspired practices in a bit.)

But first I want to digress for a moment to say that any writing program is always more than a series of goal statements, documents, or directors. It is the administrative coordinators who provide continuity across the individual directors—like Grace Crussiah and Helen McClung in the Professional Writing Program and Scott Eklund in the Academic Writing Program. And it is the teachers in the classroom who actually create the programs. That Maryland's programs have had outstanding teachers is certainly true of the first-year program, as the testimonies from directors that Scanlon gathered indicate. And the Professional Writing Program has had the overwhelming advantage of local circumstances because, thanks to Maryland's location in College Park, the program has been able to hire instructors with credentials as professional writers and trainers for the government and other institutions clustered near the nation's capital. By way of celebrating the instructors of both programs, I want to focus for a moment on Sue Oswald, a colleague who passed away suddenly twenty years ago. Sue actually began in the Professional Writing Program as a student worker (and I vividly remember her helping us frantically assemble the new instructor's manuals for a staff meeting in 1983). After graduating in English, Sue went to work for Professional Management Associates, one of the Beltway firms of the 1980s consulting on government contracts, and as a project manager there, she once proposed a joint venture with faculty of the Junior Writing Program. Coming back to Maryland as a graduate student, she was hired to teach in the program, and with her colleague and close friend Mary Scheltema—who once said unforgettably that the "currency of respect is currency"—Sue worked tirelessly to improve the contracts and pay of instructors. She was a force to be reckoned with. The best institutional circumstances and the fullest and most theoretically sound program mean nothing and create nothing without the "above and beyond" dedication of teachers like Sue.

Taking a Very Long View: The Academics of the Sixteenth Century

Maryland's academically general first-year course and its professionally oriented upper-division course look like they split the old distinction of "liberal versus practical" between them, the first course offering academic literacy skills and the second more intense preparation in a professional field or discipline. But if, as Kenneth Burke would say, we enlarge our circumference, we can find a historical understanding of rhetoric as the ground of both practices. Jim Herrick, in his chapter for this collection, shows us the enduring value of looking back millennia. I am going to look back five hundred years, for a comparison can, in fact, be made between our writing programs and the arts courses of the universities of the sixteenth century. That too was a time of enrollment expansion, the creation of new universities, and the democratization of university-level education. Though nothing like what we mean by democratization and inclusiveness today, since this was education solely for young men, access did cross social strata (to miners, blacksmiths, and even paupers) and was surprisingly international, as students, especially in the higher faculties, crossed borders and even religious divisions to attend schools outside their native countries.

Sixteenth-century universities were not organized all that differently from universities today in the sense that entering students were expected to study a core curriculum. They continued their secondary-school immersion in grammar and rhetoric and some mathematics (music and astronomy were mathematical subjects), and they added study in dialectic. Most left without degrees or went into bureaucratic careers or inherited positions; some went on to the specialized higher faculties in the professions: medicine, law, and theology (Nauert). And it is worth remembering that their courses at both the secondary and university levels were conducted in Latin and required increasingly sophisticated skills in reading and composing in this alternate language of learning. Their situation does not quite match the category of second-language, or L2, learners of today, since Latin was not the vernacular language of any living community; there were no native speakers.

Of the three fields in the trivium, the one least familiar today is dialectic. It is sometimes labeled *logic*, which creates confusions since, in the late nineteenth century, logic became symbolic logic focused on proof, leaving both probable reasoning and verbal expression behind. But sixteenth-century dialectic taught the art of inventing and judging fully expressed arguments at different levels of certainty, from the demonstrative and certain to the probable and tentative. And in the hands of humanist educational reformers, following the lead of Rudolph Agricola, dialectic and rhetoric were treated as overlapping arts. We find that overlapping in the rhetorical and dialectical treatises of Philip Melanchthon, the sixteenth-century German reformer whose pedagogy was widely followed in northern Europe. In his textbooks on both arts in the 1540s, he even expanded Aristotle's three genres, the epideictic, deliberative, and forensic, to include a fourth, the didactic, and he defined dialectic as the art of this rhetorical didactic genre by emphasizing its role in assembling and expanding the material taught in any subject area (Melanchthon, *Elementa* 422–24). In other words, Melanchthon's dialectic is Kinneavy's art of argumentation in the disciplines and, at the same time, the art of developing value-assessing and action-choosing arguments as needed in practical affairs. The list of twenty-eight topics from Philip Melanchthon's 1547 dialectical treatise shows the sources of argument in both directions (*Erotemata* 663).[4] Melanchthon's list equals the number that Aristotle has in book 2, chapter 23 of the *Rhetoric*, but it is a different set from both Aristotle's or Cicero's in the *Topica*. It is unlikely that any current writing program uses an inventional system as rich as this one from the sixteenth century. But in the actual discussion of these topics, Melanchthon—like all the other dialecticians of his time—groups them into larger categories, and these more general categories were brought into the twentieth century in Corbett's *Classical Rhetoric for the Modern Student*: definition, cause and effect, circumstance, comparison, and testimony/authority (110–46, 155–62).

Melanchthon illustrated the use of dialectic for epistemic knowledge-forming arguments with examples of the kinds of issues debated among scholars in his day (LaFontaine). To focus just on what we would now call scientific subjects, he included arguments concerning the nature of comets, the sources of earthquakes, and, with more than a passing knowledge of the medicine of his day, he gave as examples of causal reasoning the arguments over sources of pleurisy and insomnia as well as Vesalius's new theory on the origin of the veins. And coming just four years after the publication of Copernicus's *De revolutionibus* in 1543, he also set, as an example among the topics for debatable claims, whether the sun moves or whether, as in Copernicus's model, it does not move (*sol movetur, sol non movetur*; in *Erotemata* 701). It is quite amazing to see this issue set as *dissoi logoi* among devices for probable arguments in 1547. In short, Melanchthon had no limited view of a combined rhetoric and dialectic as practical and necessary arts in the formation of responsible arguers, whether in the disciplines or in civic life or, in his case, even in spiritual life.

The Rhetorical Core

What and how much, if any, of this capacious rhetoric do we retain? And here I want to turn to the "deliberative" part of my chapter and suggest what some of the essentials of this "liberal and practical" rhetorical core might be. Very roughly speaking, there are three ways of teaching a writing course: by focusing on readings following a theme or subject, by attention to genres, and by making rhetoric itself the content of the course. The thematic/content course may serve important educational goals, but a course built on readings or another body of learning puts a very large polar bear in a pup tent: it requires great skill to make room for anything else like writing instruction. The genre approach dominates in many professional writing courses by having students identify and imitate the genres of their chosen professions; writers-on-the-job are trained this way so why not aspiring writers-on-the-job? But here too there has to be a balancing act to teach the higher-level skills that would allow control and change rather than only the reflex recreation of existing genres. The rhetorically based course (*all rhetoric, all the time*), in which students write on topics of their own choosing, can also be difficult to teach because it requires teachers well trained in rhetoric. But in any of these versions, as in the rhetorically based course itself, what, in the spirit of Kinneavy, are some of the features of a rhetorical core? I am going to suggest five elements. There will be no surprises here.

1. A rhetorical approach means attention to audience—but more than that—an awareness that different audiences and occasions require different forms of accommodation. This concept of varying the selected means according to audience and occasion, the core of rhetoric, has to be a conscious element in the course. And I take audience in both the constructed and real senses, using Ede and Lunsford's durable distinction between invoked and addressed audiences. (It is possible to construct an audience textually that addresses no single plausible reader. For examples see any Wikipedia entry on a controversial technical subject.)

2. A rhetorical emphasis means an awareness that all discourse is "making a case" whether another case is present or not. There is no argument-free zone. (This view seems widely accepted.)

3. And a concentration on argument also means appreciating that cases are made with different degrees of certainty or confidence, as in the legal model of probable cause, preponderance of evidence, guilt beyond a reasonable doubt, and—only in a closed axiomatic system—proof. An emphasis on personal writing makes this understanding difficult to teach because a first-person

perspective is not externally accountable. And visuals and other multimodal appeals are difficult to hedge.

4. Some attention should be given to methods of rhetorical invention. These include the stases, as a way of identifying what is at issue in an existing argument field, and the topoi, at whatever level of richness, to assist the development of lines of argument. And now we have to include invention methods for visual or overall multimodal forms of arguing.

5. Finally, the rhetorical core includes a deep appreciation and some technical understanding of language as a set of positive options at the word, sentence, passage, and whole-text level. I am emphasizing language here, but this element holds for any semiotic system. The operative word here is *positive*, teaching functional options, whether from discourse analysis like given/new coherence strategies or from the older rhetorical stylistics like loose versus periodic sentence structure. In item 5 here we return full circle to item 1, since accommodating different audiences means an awareness of the different formal means of doing so. Teaching all five core elements requires more than one course can easily deliver, but I nevertheless want to suggest briefly what attention might mean in the classroom for two of these core features: *accommodation* from the first and *amplification* from the last.

The Rhetorical Core: Accommodation and Amplification

Accommodation: Answering the Problem of Expertise

Let's go back to the opening reminder of the current emphasis on STEM education––and to another current problem traced to it—the increasing economic divide in the United States and, underlying it, the increasing divide in expertise between a few who have a monopoly on the knowledge that drives the economy and makes the world work and the majority who are more and more subject to this esoteric knowledge and less and less capable of informed decision-making (Rotman). To address this divide, all writing courses should teach audience adaptation/construction, but professional writing courses should emphasize the skills of mediation, the accommodation of expertise to nonexpert audiences who often are asked to support or adopt the recommendations of experts and live with the consequences, good or bad. For the discourses of science, I prefer the term *accommodation* to the term *popularize* because the former suggests multiple audiences and mutual adjustments. As Kinneavy suggested, while they become specialists learning to communicate with each other, professional writing students should also learn to be generalists, able to adjust their positions persuasively and empathetically to what other publics might need to know ("Writing" 17–18).

This emphasis on accommodation is visible in the first manual of the Junior (now Professional) Writing Program (a manual with the thoroughness that only Michael Marcuse could bring to a document, covering everything from rhetorical theory to parking stickers). The original statement of goals and standards for the program specified having students write to varying audiences on subjects associated with their intended majors and careers:

> The audience is not the already-informed but rather readers with utilitarian purposes, the more typical audience for professional communications. . . . The Professional Writing program mediates between a student's general education and a student's specialized studies, and between a student's specialized discourse and his or her general community. . . . [S]tudents learn to build verbal bridges between their fields of knowledge and the minds of uninitiated non-specialists upon whom that knowledge impinges. (Marcuse 1.B)

Sounds good, but how to do it? For one, by having students address real audiences as much as possible, as in current service-learning courses. We took real audiences so seriously in the early days of the Professional Writing Program that we had students turn in their final proposal arguments with an addressed, stamped envelope for mailing them; it is much easier today to post them on a website. To develop the skills of accommodation per se, at least one assignment in any course should ask students to address the same argument to different audiences—with the understanding that the case will never remain quite the same.

We experimented with such assignments like the "definition for two audiences." Students were encouraged to take a concept they were investigating in a current course and adapt it to two audiences with different levels of expertise. This assignment allowed us to focus on adjustments in the language and arrangement for different audiences so that we could ask difficult questions about what survived from one version to another. In trying to teach this assignment, we needed published examples where the same case was addressed to audiences at two levels of expertise or commitment (and that led me personally to a lifelong scholarly interest). But this assignment turned out to be a devil to fulfill——because it asked students who were novices in their majors to imagine themselves as experts and because it brought the problem of exigence front and center—the need for an "argument for the argument," a reason for addressing a particular audience on the material in the first place. The harder of the two was always the more accommodated, more "outsider" piece, because it required this construction of exigence usually taken for granted within a discipline and because it called for the opposite of what was taken to be the stylistic standard in professional and technical writing: brevity.

Amplification: The Opposite of Brevity

Brevity as the source of clarity is an often-lauded stylistic goal in professional writing. Yet the assumption that the shorter the text, the clearer it is, *necessarily*, is not found in the rhetorical tradition. Clarity as understandability is certainly recommended in rhetorical treatises as one of the four virtues of style (correctness, clarity, appropriateness, force), but it is not equated with brevity. (*Brachylogia*, a series of short sentences for emphasis and prosodic effects, is recommended, as are compact single-sentence definitions for summative force.) The current strength of the clarity-equals-brevity notion may have roots in the efficiency experts of the early twentieth century and in an attitude derived from the logical positivists who regarded non-fact-like assertions as nonsense. It may also stem from a justified distrust and fear of bureaucratic obfuscation epitomized in George Orwell's 1930s polemic "Politics and the English Language." The journalistic version of clarity equals brevity can also be found in the many midcentury readability formulas like those of the apostle of journalistic style, Rudolf Flesch, or in the Gunning Fog Index. And it is enshrined in Strunk and White's influential *Elements of Style*. There are two enduring legacies of this emphasis on brevity: an equation of average sentence length with readability and grade-level accessibility (even the Nancy Drew mysteries were rewritten) and an obsession with cutting out the "deadwood" in professional writing, an exercise often used in training on-the-job writers.

But, in fact, rhetoric texts from Cicero on recommend the opposite stylistic standard of *amplification*.[5] Stylistically, to amplify a text means to open it up for audience-sensitive reinforcement, so teaching amplification requires a language/content inventiveness natural to the rhetoric classroom. In an amplified understanding of amplification, the concept comes in two senses in answer to the question—how do you make something more important? The first answer involves devices for heightening or highlighting an element in a text. The second involves devices for increasing the space or "presence" that an element has (Fahnestock 390–417). To explain this difference with a rough visual analogy, heightening comes from variations in focus or lighting or color variation for any element in a visual field (tactics taught in visual rhetoric), and extension comes from filling the viewer's field of vision to the exclusion of anything else.

How are these effects achieved in language? Rhetoricians from antiquity onward had methods for doing both verbally. For instance, you can make an element stand out, without taking up more space, by using a register shift, creating "pop" by borrowing from another semantic or usage field. And you can make an element take up more space by inserting reinforcing material that keeps the reader's attention on it—often in the form of comparisons or illustrations or added descriptive or narrative detail. After discussing these principles and seeing them in action, students can practice them in two ways. First, they can take a short text and identify where and how to expand it. This exercise can use a found text that is short (as easily found in *Scientific American* or *Technology Review*), or a shortened version of a longer piece can be created and students can compare their proposed expansions to what the author actually wrote. It is better when students do these "riffs" across subject matters in a rhetoric-based course that mixes students with different majors; that way the amplification options become extractable, general skills. The skills of amplification then feed directly into revision assignments. Second, students can be asked to revise a paper through heightening and extending, usually to a specified length, and then to explain what they did in an accompanying memo on why the chosen strategy would help in a particular accommodation scenario. Even better, students can project their own work or someone else's on a classroom screen and discuss what they tried and how. Experienced teachers will also appreciate the excitement generated when small groups, or the whole class together, do these expansions on the spot, collaboratively.

The Future Discipline: The Growth of Professional Writing Programs

These are just two classroom tactics delivering on two of the five items in the rhetorical core. There is more than enough similar material to cover on these issues and others in coordinated academic and professional writing courses like those at Maryland, and now at many, many other schools. And that brings me to the "future" part of my title. With a few notable exceptions like Carnegie Mellon and New York University, professional writing programs like Maryland's did not really exist forty years ago. But now we are in boom times in the expansion of professional writing programs that not only supervise undergraduate courses and concentrations but also offer certificates and graduate degrees. To reach their far-flung, on-the-job constituents, many of these programs, especially at the master's level, are offered online. The recent growth in professional writing programs is a response to both the current STEM emphasis and to the need for multimodal composing skills given our current communication media. This growth matches the earlier expansion of graduate programs and concentrations in rhetoric and composition in the 1980s and 90s, which created undergraduate and graduate programs still housed in English departments.

In addition to these writing programs, professional and general, within traditional English departments, there are now also many more independent writing programs outside departments, with some like the one at the University of California, Davis, functioning like departments. And a kind of leading edge of this expansion is the recent trend in creating stand-alone departments of rhetoric and writing. Some departments, like those at Minnesota, Iowa, and Berkeley, have been around for a century. But most of the more than thirty other departments of rhetoric and writing studies now existing have appeared within the last twenty years, and many more recently than that: those at the University of Utah and the University of Central Florida are new this year. The creation of new departments requires the creation of new courses and new degrees and fosters a new body of learning, just as Jonathan Turner predicted over one hundred and fifty years ago.

But we need not stay within national borders to see the expansion and growth of academic and professional writing programs. The WAC Clearinghouse makes available online a published anthology of program descriptions called *Writing Programs Worldwide*. Universities abroad now also have their own rhetoric and writing departments, like the University of Amsterdam in the Netherlands and Lund University in Sweden. And drawing on Western and non-Western rhetorics, the first department of rhetoric and communication opened in a Chinese university (Beihang University) in 2011. Still another sign of this global growth in programs and departments is the appearance of new professional organizations. Just as there has been a Rhetoric Society of America since the 1980s, there is now a much larger Chinese Rhetoric Society and Global Rhetoric Society (including the Japanese and Korean Rhetoric Societies). We have the beginnings of a Sociedad Latinoamericana de Retórica, and within the last few years, a new Rhetoric Society of Europe and Rhetoric Africa (centered at the University of Cape Town) have appeared, all sponsoring meetings and publications. These initiatives around the globe, under the rubric of rhetoric, have varying emphases. But clearly the *once* is now the *current* and arguably the *future* discipline, and programs in both academic and professional writing in U.S. schools now stand to benefit from worldwide perspectives.

I have made this closing argument for the future growth of rhetoric and composition courses, programs, departments, and societies by using the kind of advice I would learn in a rhetorically based writing course, drawing on the topics of Aristotle listed in book 2, chapter 19 for predicting future outcomes from current trends (Roberts 132):

That a thing will be done if the person is actually setting about it
That a thing will be done if there is both the power and the wish to do it
That if there is a foundation, there will be a house.

Notes

1. The yearly catalogs spanning the history of Maryland Agricultural College, then Maryland State College, and later University of Maryland are available in an online archive at http://www.lib.umd.edu/univarchives/catalogs.

2. See the University of Maryland timeline, http://www.umd.edu/timeline/.

3. In effect, Maryland simply moved a second semester requirement, traditionally filled by a course in literature, to the junior level, but in doing so it created a completely separate writing program and a novel experiment for its time. In its first conception, Maryland's program was to use instructors from different departments to teach writing courses tailored to their majors. But absent willing faculty, or any structures or faculty in place to train them, other departments quickly turned staffing over to the new program. The requirement itself was trimmed somewhat by an exemption releasing any student who earned an A in freshman writing at Maryland, or in a similar first-year program, from the upper-division requirement—part of the "one-size-fits-all" view again. (The College of Engineering never used this exemption.) Meanwhile the freshman writing program had its own healthy exemption based on SAT scores and Advanced Placement credit. However, in the revision of the general education requirements in 2010, all these exemption structures were removed with the exception of Advanced Placement credit for a score of 4 or 5 on the English composition test as the equivalent of English 101. There is no exemption from the Professional Writing Program, which now offers more than fifteen different courses (for example, writing for nongovernmental organizations and writing in the sciences) compared with the original two: advanced composition and technical writing.

4. Melanchthon's topics are as follows: definition; genus; species; differentia; etymology; conjugates; whole–parts; division; causes; effects; antecedents; consequence; from the absurd; from the necessary; from the impossible; adjuncts; circumstances; common accidents; similars; equals; from the greater; from the lesser; proportion; opposites; disparates; signs; examples; authority.

5. On amplification see Cicero's *De inventione* for ways of expanding syllogisms (100–03) and Quintilian, *Institutio oratoria*, book 8, chapter 3, for a general review of tactics. Brevity was defined as the opposite of amplification. In the Renaissance, Erasmus's *On Copia* offered tactics for amplification in both language and content.

Works Cited

Berlin, James. *Rhetoric and Reality: Writing Instruction in American Colleges, 1900–1985.* Southern Illinois UP, 1987.

Brereton, John. "Liberal Arts vs. Vocationalism: A New Perspective." Conference on College Composition and Communication, 2003. Unpublished talk.

———. *The Origins of Composition Studies in the American College, 1875–1925.* U of Pittsburgh P, 1996.

Brown, Donald R. "Jonathan Baldwin Turner and the Land-Grant Idea." *Journal of the Illinois State Historical Society (1908–1984)*, vol. 55, no. 4, 1962, pp. 370–84.

Cicero. *De invention; De optimo genere oratorum; Topica*. Translated by H. M. Hubbell, Harvard UP, 1976.

Corbett, Edward P. J. *Classical Rhetoric for the Modern Student*. 2nd ed., Oxford UP, 1971.

Ede, Lisa, and Andrea Lunsford. "Audience Addressed/Audience Invoked: The Role of Audience in Composition Theory and Pedagogy." *College Composition and Communication*, vol. 35, no. 2, 1984, pp. 155–71.

Erasmus. *On Copia of Words and Ideas*. Translated by Donald B. King and H. David Rix, Marquette UP, 1963.

Fahnestock, Jeanne. *Rhetorical Style: The Uses of Language in Persuasion*. Oxford UP, 2011.

Fahnestock, Jeanne, and Marie Secor. *A Rhetoric of Argument*. McGraw-Hill, 2004.

Fleming, David. *From Form to Meaning: Freshman Composition and the Long Sixties, 1957–1974*. U of Pittsburgh P, 2011.

Kinneavy, James. "Restoring the Humanities: The Return of Rhetoric from Exile." *The Rhetorical Tradition and Modern Writing*, edited by James J. Murphy, Modern Language Association, 1982, pp. 19–30.

———. "Writing across the Curriculum." *Association of Departments of English Bulletin*, vol. 76, Winter 1983, pp. 14–21.

Kitzhaber, Albert. *Themes, Theories, and Therapy: The Teaching of Writing in College. The Report of the Dartmouth Study of Student Writing*. McGraw-Hill, 1963.

LaFontaine, Mary Joseph. *A Critical Translation of Philip Melanchthon's* Elementorum Rhetorices Libri Duo [Latin text with English translation and notes]. U of Michigan, PhD dissertation, 1968. Microfilm.

Marcuse, Michael. "The History of the Maryland Professional Writing Program." *Instructor's Manual: Professional Writing Program*. U of Maryland, 1983.

Melanchthon, Philip. *Elementa rhetorices*. In *Opera quae supersunt omnia*, edited by C. G. Bretschneider, vol. 13, Johnson Reprint Corporation, 1963, pp. 415–506.

———. *Erotemata dialectices*. In *Opera quae supersunt omnia*, edited by C. G. Bretschneider, vol. 13, Johnson Reprint Corporation, 1963, pp. 509–752.

Mervis, Jeffrey. "Science Standards Begin Long, Hard Road to Classroom." *Science*, vol. 340, no. 6139, 21 June 2013, p. 1391.

Morrill, Justin S. *Speech of Hon. Justin S. Morrill, of Vermont, on the Bill Granting Lands for Agricultural Colleges: Delivered in the House of Representatives, April 20, 1858*. Washington, DC, Congressional Globe Office, 1858.

Morrill Act of 1862. *www.ourdocuments.gov*, www.ourdocuments.gov/doc.php?flash =true&doc=33.

Nauert, Charles G. "Humanist Infiltration into the Academic World: Some Studies of Northern Universities." *Renaissance Quarterly*, vol. 43, no. 4, 1990, pp. 799–812.

Nelson, Bonnie. *Freshman English at Nine Institutions Which Have Eliminated the Traditional Course in Freshman Composition: Antioch, Elmira, Juniata, and Swarthmore Colleges and Baker, Clark, Emory, Maryland, and Tulane Universities*. Modern Language Association, 1968 [Eric TE 500 190].

Roberts, W. Rhys, translator. *The Rhetoric and the Poetics of Aristotle*. Modern Library, 1984.

Rotman, David. "Technology and Inequality." *Technology Review*, vol. 117, no. 6, 2014, pp. 52–60.

Scanlon, Molly J. *A History of the Introductory Writing Program at the University of Maryland (1958–1978)*. U of Maryland, MA thesis, 2008.

Stage, E. K., et al. "Opportunities and Challenges in Next Generation Standards." *Science*, vol. 340, 19 Apr. 2013, pp. 276–77.

Turner, J[onathan]. B. *Industrial Universities for the People. Published in Compliance with Resolutions of the Chicago and Springfield Conventions, and under the Industrial League of Illinois*. Jacksonville, IL, Morgan, 1853.

Venn, John. "On the Diagrammatic and Mechanical Representation of Propositions and Reasonings. *The London, Edinburgh, and Dublin Philosophical Magazine and Journal of Science*, vol. 10, no. 5, July–Dec. 1880, pp. 1–18.

Section II

Programmatic Perspectives at Illinois, Cornell, Howard, and Penn State

3

Undergraduate Rhetoric at UIUC: Revising a Curriculum, Rethinking a Program

Kelly Ritter

Providing a snapshot of the Undergraduate Rhetoric Program at the University of Illinois at Urbana-Champaign as of fall 2014, Ritter discusses the pedagogical and theoretical principles upon which the program operates, its storied history, its evolving student body—with three diverging populations of traditional majority students, underrepresented students, and international students—and the rhetoric courses they take. She closes by identifying four key challenges common to nearly all large writing programs at institutions nationwide: shrinking graduate enrollments and an increasing number of undercompensated non-tenure-track positions, inadequate preparation in the teaching of writing, the pervasiveness of dual-credit programs, and the surge of international students and multilingual learners.

Our Undergraduate Rhetoric Program (first-year writing) at the University of Illinois at Urbana-Champaign (UIUC) has a prominent place in the historical record of rhetoric and composition studies. We are mentioned numerous times in the archival documents of late nineteenth- and early twentieth-century writing instruction featured in John Brereton's *Origins of Composition Studies*. Our basic writing program's political and curricular history is the subject of Steven Lamos's monograph *Interests and Opportunities*. Our UIUC writing center (in its earliest formations) features prominently in one of Neal Lerner's historical works ("Searching for Robert Moore"). And one of our early Undergraduate Rhetoric Program directors, Professor Charles Roberts, was the first editor of the journal *College Composition and Communication*. Beyond this prominent history, we are joined in space and intellectual spirit in our English building by the Center for Writing Studies, which began in 1990 under Professor Gail Hawisher and continues to serve as an exemplar of interdisciplinary graduate writing and writing across the curriculum to this day. And we were the recipient of a 2011–2012 Certificate of Excellence for Writing Programs awarded by the Conference on College Composition and Communication.

This storied history and notable position in the network of writing programs nationwide brings with it both responsibilities and challenges. What I attempt to do in this overview is outline who and what the UIUC Undergraduate Rhetoric Program is *now*, being mindful of how important such declarations of identity and intent can be. Articulating who we are now includes spelling out how we operate—our primary goals and desired student learning outcomes—and in what ways the future brings ever-new considerations for our program's faculty, students, and staff. Archival snapshots of first-year writing programs for the sake of documenting practices and resource allocations are few and far between; more deliberate accounting of this kind would certainly aid scholars of writing studies in tracking the progress of the first-year course on a local and regional level. The most recent in-print and "official" snapshot of the UIUC Rhetoric Program appears in the 1993 *Profiles of Writing Programs*, published by the Alliance for Undergraduate Education (Working Group on Writing Instruction). As we look at the UIUC Rhetoric Program more than twenty years later, some things have changed dramatically, while—as the saying goes—other things have stayed remarkably the same. As I will outline below, through answers to a series of broad questions, our student body (and enrollment size) has dramatically shifted, while our course structure has remained relatively stable (with the exception of the now-defunct Computer Rhetoric 108 course offered in 1993). We have a larger programmatic staff that includes graduate students and supports the ever-professionalized discipline of rhetoric and composition within the university. We also now have a large and valuable non-tenure-track faculty teaching in our program and working alongside our graduate teaching assistants (TAs)—a condition not in place in 1993, at UIUC or at many other universities. With these changes comes more attention to the importance (and acceptance) of an evolutionary process—both forward and potentially backward—that is endemic to any writing program, but particularly one of our size.

Our UIUC Rhetoric Program—which includes basic, standard, and advanced composition courses under its umbrella—aims to deliver valuable, rigorous courses to thousands of first-year students from across the disciplines, like so many other programs at other large institutions, while serving an exceptionally well prepared student body that increasingly arrives to us from outside the United States. We are a global campus, and so we are a global program. The Illinois of yesterday, therefore, is not necessarily the Illinois of today, even as some statistics remain constant. As director of the program (at the time of the 2014 Maryland Conference), I must therefore balance histories and futures as I keep always in view some questions: What *is* the Rhetoric Program at Illinois; what does a student of rhetoric need on our campus; and how can (or should) we meet that need?

Who Is the UIUC Rhetoric Program?

At UIUC we operate a very large program—on average, providing about a hundred sections of rhetoric courses per semester, all taught by graduate students and non-tenure-track faculty (who number, overall, one hundred graduate students and twenty-nine non-tenure-track faculty). Many of these instructors also teach introductory or advanced literature, film, and creative writing. We are very fortunate to have the resources that provide a full office staff and support system for these instructors and their students, as well as for me as director. Our Rhetoric Program office consists of a tenured writing program administrator (WPA), who receives one course release per semester for her work, plus a tenth month of salary and a modest annual stipend. We also have a full-time associate director at the rank of academic professional (a non-tenure-track position that carries with it reasonable security of employment), who works in the office forty hours per week. In addition, our office includes two graduate student assistant directors, typically doctoral students in English (often in the writing studies concentration), who work at 67 percent time, or approximately twenty-six hours per week. These assistant directors work closely with the WPA on curricular issues, TA orientation, new TA training, and the twice-annual Undergraduate Rhetoric Conference, which showcases student work from our courses (both in panel presentations and poster sessions) in a professional setting. And, most critically, we have a full-time office administrator (staff), who also works forty hours per week and assists both the Rhetoric Program and the associate head of the English department with scheduling, student concerns and complaints, hiring paperwork, graduate student teaching assignments, and other short- and long-term program issues.

Beyond these core faculty and staff, we employ four to six graduate student peer mentors each fall semester, at 33 percent time (or approximately thirteen hours per week). These graduate students work with small groups of new TAs—typically four to five on average—on their teaching of our first-year course Rhetoric 105: Writing and Research. The mentors meet weekly with the new TAs and assist with lesson planning, issues in grading and evaluation, classroom management, and other course development questions. Their work augments English 593: Proseminar in the Teaching of Rhetoric that all new TAs take in their first semester, while teaching a section of Rhetoric 105. The

WPA often teaches the 593 course (our associate director and I take turns as lead instructor for the course). We also employ a digital literacies coordinator at 33 percent time; this graduate student helps maintain our program website and our course databases (including suggested readings, sample syllabi and assignments, and other resources) and also assists instructors with technology needs inside and outside the classroom. She also inventories and maintains our Macintosh laptops, checked out by instructors for in-class use as needed. Finally, an additional staffing pilot that we offered in 2014–2015 was for two peer development coordinator positions, filled by non-tenure-track faculty (or "specialized faculty," as our newest institutional nomenclature deems these instructors), who were charged with providing resources and mentoring for the approximately twenty-nine non-tenure-track faculty teaching with us and with making these instructors feel like a larger part of our Rhetoric Program as a whole. These peer development coordinators created a non-tenure-track faculty lecture series, began reading and discussion groups for these faculty members, and made themselves available for peer observations of teaching. We hope that, funding willing, this pilot project will continue into subsequent years.

Who Are UIUC Students?

As I mentioned in the introduction, we have a large and diverse student body at UIUC, one that reflects the global economy in which we live and work. For the class of 2017 (those students who matriculated in fall 2013), the statistics are as follows:

- 33,201 students applied, 7,331 students admitted
- average ACT score: 28.6
- 54.5 percent in the top 10 percent of their high school graduating class
- 22.0 percent are first-generation college students
- average age: 18.7 years
- 56 percent male, 44 percent female
- 16.1 percent are from underrepresented/minority groups (33 percent of whom are Asian or Asian American students)
- 72.5 percent are residents of the state of Illinois
- 16 percent are international students (with permanent residence addresses outside the United States)

As I emphasized in my introduction, our student population is well prepared and significantly non-U.S. born. To further illustrate the latter of these two conditions, figure 3.1 graphs the growth of international students on our campus between fall 2000 and fall 2012. As we state in our *Rhetoric Instructors Handbook*—which all new TAs and non-tenure-track faculty receive the first time they teach in our program:

> since 2008, the undergraduate international student population at Illinois has more than doubled—rising from 2,232 to 4,996 students. Much of this growth has been driven by rising numbers of Chinese international undergraduates, whose population grew from 258 in 2008 to 2,588 in 2014. The university historically has also had a large South Korean international student population. Currently, Illinois has the largest international student population of *any* public university in the United States. (*Rhetoric* 8)

Given this reality, we see in our rhetoric classrooms a mix of predominantly white, resident students of the state of Illinois (significantly hailing from what is known locally as the "Chicagoland" area—that is, the city and its many suburbs) and international students, with (as the class of 2017 statistics reveal) a smaller percentage of minority and underrepresented students. One of our ongoing challenges, therefore, is how to best serve these three populations—and how to aggressively recruit and retain students from underrepresented groups who are not also from international admissions so as to diversify our student body further. We have courses in place in the linguistics department for international, English as a second language (ESL) students who score on the lower end of the TOEFL (Test of English as a Foreign Language) and

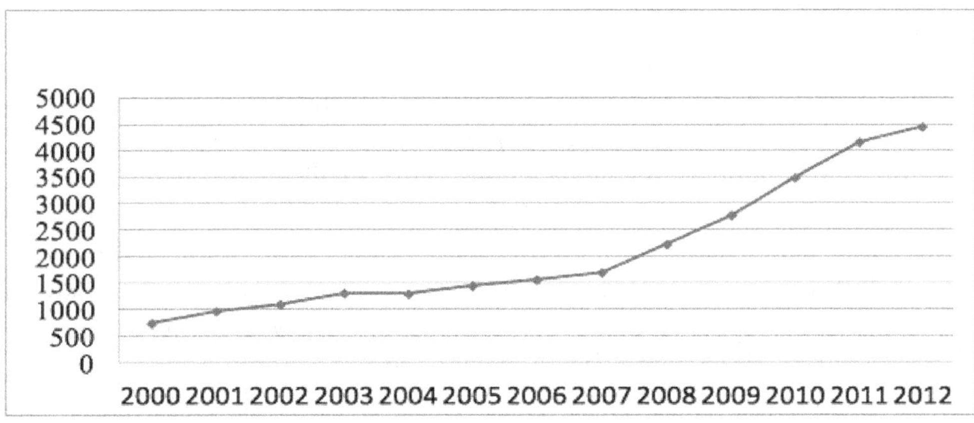

Figure 3.1. Number of international undergraduate students at the University of Illinois at Urbana-Champaign, by year. The undergraduate student body in 2012 numbered 31,901. Rhetoric Instructors Handbook, p. 8.

local English placement exams and who need more sustained help with writing (see below for more on this), but as readers know, to be a second-language learner is not always to be in need of a separate course focusing on language acquisition, and writing programs often privilege U.S.-centric versions of writing and writers (Matsuda).

So, our instructors and our program staff are ever mindful of the needs of these second-language learners, high-achieving and highly motivated international students who enroll in our rhetoric sections alongside native-speaking, largely in-state white students who typically also come to the university with high test scores, top high school grade point averages, and accompanying expectations of scholastic achievement. Expectations from both student groups are, unfortunately, that the rhetoric class is *less* challenging than it actually is (a view that first-year writing is "remedial" as compared with their overall preparation and institutional positioning). This leads to concomitant expectations (or hopes?) of less work for higher grades and some resentment over placement, particularly among native-speaker students. This is a challenge we face in designing and delivering our courses—a challenge not necessarily unique to UIUC but a persistent one nonetheless.

What Does Our Rhetoric Program Offer?

Currently our program offers four writing courses, three designed for first-year writers and one designed for those beyond the first year: Rhetoric 101: Principles of Writing; Rhetoric 102: Principles of Research; Rhetoric 105: Writing and Research; and Rhetoric 233: Advanced Rhetoric and Composition. Rhetoric 101 and 102 are designed as a sequence, and they form our basic writing curriculum (though our catalog descriptions or other published literature on these courses does not designate them as "basic"). A student with an ACT English subscore of 19 or below is currently placed into the Rhetoric 101–102 sequence, which includes a weekly conference with the instructor, a tutorial of thirty to sixty minutes. A student with an ACT English subscore of 20–31 is placed into Rhetoric 105. Students who receive a 32 or higher are exempted from the course requirement, Composition I. Both Rhetoric 101–102 and Rhetoric 105 fulfill this Composition I requirement for graduation, which is standard across all majors and all programs and colleges at UIUC. Rhetoric 233, in contrast, is an elective on UIUC's menu of advanced composition courses, which are designed to fulfill the upper-level writing requirement across the university, or the Composition II requirement as it is sometimes called.

Rhetoric 101–102 is capped at sixteen students per section, and Rhetoric 105 is capped at nineteen students per section, bringing us within the guidelines of the National Council of Teachers of English (NCTE) for total number of composition students assigned to a given instructor per class. (Our TAs teach no more than two sections, and our non-tenure-track faculty teach no more than three). Of course, ideally, our sections would be smaller; my personal desire would be to cap Rhetoric 101–102 at twelve and 105 at fifteen. Readers know that the professional guidelines of both the NCTE and the Modern Language Association promote a maximum of fifteen students per section of basic writing and a maximum of twenty students per standard composition section, with no more than forty-five to sixty students per instructor, per term (ADE Guidelines). Our guidelines follow the spirit (and nearly the letter) of these national guidelines. Readers further know that the arguments for smaller class size in writing courses are well rehearsed, if slightly inconclusive in terms of specific numbers or measurable data (see Horning, "The Definitive Article on Class Size," and more recently, the NCTE's "Why Class Size Matters Today").

In terms of placement, about 100–150 students take the 101–102 sequence annually, and on average, 3,000 students—or about 50 percent of an average incoming class—are exempted from the Composition I requirement based on their ACT subscore (31 or higher), their Advanced Placement literature or language exam score (4 or 5), a requisite International Baccalaureate score, equivalent transfer course credit from another institution, or, theoretically, a dual-credit program taken in their high school. Most of the remaining students (~2000–2500) take Rhetoric 105, although there is a cognate sequence of courses in the Department of Communication (Communication 111–112, a two-course sequence in speech communication) for those who do not want to take rhetoric; this option was elected by 800 students in 2013–2014. We also have the aforementioned course in the Department of Linguistics (ESL 111–112, also a two-course sequence) for students whose TOEFL scores indicate a need for further testing, done by that department, the results of which determine whether a student will take the ESL sequence. Approximately 650 additional students enrolled in this ESL course sequence in 2013–2014. This sorting mechanism for international students (stateside second-language students typically are not part of this testing group) began several years ago, and it allowed for more concentrated attention on a population in need of additional help with writing.

The segregation of these students into courses outside the rhetoric sequence, however, is something that needs more research, in terms of its social implications alone—if one sees the first-year writing course as an important space for building community and establishing a culture of writing among students. This also applies to the significant number of students who test out of the Composition I requirement altogether, which is a vexing issue to me and many other WPAs whose institutions enroll significant numbers of students qualifying for Advanced Placement, early college, or dual-credit exemptions upon admission. A fuller discussion of these issues critical to higher education today, and particularly to general education courses like first-year writing, can be found in Hansen and Farris, Addison and McGee, and the NCTE's "Statement on Dual Credit. These resources, which I draw on in my administrative work and in training our graduate students in the office (and in which I participated, for the very early working-group stages of the NCTE position statement), highlight important research on the conditions and effects of dual credit / dual enrollment for high school students and first-year college programs.

Rhetoric 233, designed primarily for the Composition II requirement (especially for students whose major fields do not offer writing-intensive courses that fulfill this requirement), is also taken by transfer students who may have a course that meets some, but not all, of the components of our Rhetoric 105 course and thus are in need of completing this requirement. Rhetoric 233 is capped at a comparatively large twenty-four students per section, as this is the standard course cap for all advanced composition course offerings across the university. This course is typically themed in nature and has, in recent semesters, included a multimodal component in many sections, as we aim here to introduce students—many of whom are sophomores or juniors—to varied means of composing beyond those introduced in our first-year rhetoric sections. An ongoing challenge is the mix of first-year transfers with second- and third-year continuing students in the Rhetoric 233 classroom. Because we offer a four-credit rhetoric course at the first-year level, many students coming from three-credit course structures fail to meet the complete requirements for transfer. Currently we are seeking alternatives to this catchall approach for transfer students who fall into this category. Certainly, as we refine and shape our student learning outcomes, or SLOs, for this course (see below), the unpredictable mix in student enrollment for Rhetoric 233 is in the forefront of our minds.

In terms of the *content* of our rhetoric courses, we have both localized and generalized goals in mind. Within our Rhetoric 105 course sections, we offer specially dedicated sections for living-learning communities on campus. These are the Global Crossroads program (which emphasizes global and international concerns in its curriculum), the Weston Exploration program (designed for liberal studies majors), the Ethnography of the University program, and the

Unit 1 group (a more general living-learning community housed in one of our dormitories, Allen Hall). Across each of our Rhetoric 105 sections, however, as well as our Rhetoric 101–102 sequence and our Rhetoric 233 sections, we have standard SLOs for each course that attempt to offer flexibility in specific course design (assignments, readings, activities) while still maintaining standards across sections that will guide instructors toward common goals for student achievement and program continuity. As a WPA, I find this mix of commonality and flexibility critical in administering a program as large and diverse as ours, and so we created these SLOs in 2013–2014 with full implementation in fall 2014. Prior to this date, our program generally operated under the auspices of the WPA Outcomes Statement (it was included in full in our handbook for new instructors), but these outcomes were neither enforced nor explicitly tied to any syllabi or writing assignments. Following principles found in both the NCTE's "Principles for the Postsecondary Teaching of Writing" and the conditions necessary for writing, reading, and critical analysis from the *Framework for Success in Postsecondary Writing*, we collectively crafted the following SLOs for each course, with input from our rhetoric instructors, past and present, and with consideration of the values and pedagogies already at work in our courses.

After completing Rhetoric 101: Principles of Writing,
students will be able to:

1. Distinguish between the conventions of academic and nonacademic texts (print and/or multimodal).
2. Summarize, interpret, and evaluate arguments found in nonfiction texts (print and/or multimodal).
3. Compose argument-driven texts that respond to exigent issues, problems, or debates.
4. Reframe their writing in response to different rhetorical situations.
5. Describe and reflect on their own writing processes, including revisions made in consideration of peer and/or instructor feedback.

After completing Rhetoric 102: Principles of Research,
students will be able to:

1. Identify and explain the role that rhetorical appeals and the rhetorical triangle can play in nonfiction print and/or multimodal texts.
2. Create and sustain across one or more pieces of writing a focused research question that responds to an exigent issue, problem, or debate.
3. Compose cogent, research-based arguments, in print-based and/or multimodal texts, for specialist and/or nonspecialist audiences.

4. Locate, accurately cite (through summary, paraphrasing, and quoting), and critically evaluate primary and secondary sources.
5. Demonstrate knowledge of writing as a process, including consideration of peer and/or instructor feedback, in one or more pieces of writing from initial draft to final revision.

After completing Rhetoric 105: Writing and Research,
students will be able to:

1. Identify and explain the role that rhetorical appeals and the rhetorical triangle can play in nonfiction print and/or multimodal texts.
2. Create and sustain across one or more pieces of writing a focused research question that responds to an exigent issue, problem, or debate.
3. Compose cogent, research-based arguments, in print-based and/or multimodal texts, for specialist and/or nonspecialist audiences.
4. Locate, accurately cite (through summary, paraphrasing, and quoting), and critically evaluate primary and secondary sources.
5. Demonstrate knowledge of writing as a process, including consideration of peer and/or instructor feedback, in one or more pieces of writing from initial draft to final revision.

After completing Rhetoric 233: Advanced Composition and Rhetoric,
students will be able to:

1. Evaluate the effectiveness of claims and advanced rhetorical strategies employed in complex arguments in nonfiction print and/or multimodal texts.
2. Situate their ideas in conversation with relevant discourse communities through appropriate source selection, evaluation, and integration (including proper citation practices).
3. Compose arguments in print and/or multimodal texts for a specific discourse community that synthesize multiple and/or competing perspectives.
4. Engage in writing as a recursive process that includes reflection and response to feedback and that culminates in publication within a peer community.

Readers may notice that the SLOs for Rhetoric 102 and 105 are *identical*; this is because we recently retooled our entire rhetoric curriculum to better sequence the 101 and 102 courses in relation to the 105 course. In the process we also renamed our courses to better reflect their actual content (and to eliminate arcane course descriptions that no longer accurately described what each course taught). Whereas previously the 101 and 102 courses were a "stretch" version of 105 (though not specifically designed with the so-called stretch model of basic writing, made popular at Arizona State University some years ago, in mind; see Glau for a discussion), we now have crafted a separate 101 course that offers a more deliberate expository introduction to academic writing approach against the 102 course, which is meant to be identical in scope to Rhetoric 105. This design allows us to move students toward the same end point, in terms of competencies and abilities. It also avoids the problem that happened in previous years in the 101–102 sequence when students would move from one instructor to another between semesters, but the instructors would not have coordinated where the "split" between the two courses happened. This resulted in either repeated work or lost work altogether. Having a separate set of SLOs govern 101 and 102, respectively, keeps instructors paced in a similar manner, while allowing for freedom and flexibility in individual course design. Making 102 and 105 identical courses reinforces the program's view that students placing into 101–102 are not remedial, or in need of totally segregated instruction, but simply need—for whatever reason—more writing help before taking the 105 curriculum. This system also allows us to assess 102 and 105 student work in consort in a program-wide assessment that we plan to conduct in summer 2015.

To expand on what our program values—and supplement the shorthand version in the SLOs—here are some of the guiding principles as I would articulate them:

- An academic writing course should focus on guided instruction in creating, developing, and sustaining *exigent arguments based in the principles of rhetoric.*
- Such a course should also include instruction in *crafting and executing research proposals and projects,* including responsible and meaningful engagement with sources.
- Lower-order concerns should be taught as needed but *always in the context of* higher-order concerns.
- *Writing is a process* that involves peers as well as instructors, one-on-one conferences for student and teacher, and individual and group learning opportunities.
- College writers should be guided toward understanding their work as critical to *present and future participation in the public sphere.*

Part of what affects our goals is our campus itself. I certainly believe, as a WPA and as a scholar of rhetoric and composition, that all pedagogy is local.

So, in addition to privileging universally valued elements of the writing process in our goals, and prioritizing rhetoric as a unifying concept for teaching writing and argument (a stance promoted, also, by many in our field), at UIUC we have a campus community highly invested in research, and our faculty strive to provide undergraduate research opportunities at every turn. Therefore, while research-based writing is not (at least for me) the be-all, end-all of academic writing instruction, it *does* have a central place on our campus. So, in our rhetoric courses, we want to introduce students to that process of inquiry so that they will have some foundation for its principles when moving into their major fields of study. Such transfer of knowledge and skills, as scholars in our field have shown, is certainly fraught with complications and thus currently the subject of further investigation (see Yancey et al.; Adler-Kassner and Wardle), but overall we emphasize the *inquiry* more than the strict formation of a research project. Instructors have flexibility in whether they require a traditional final research paper or require smaller pieces of inquiry-based writing through the term. Our only requirement is that the course have *some* kind of emphasis on research and inquiry across its major writing assignments. Our other requirements are that Rhetoric 105 students write at least twenty-five pages of revised prose and that 101–102 students write at least twenty pages, comparatively. (Rhetoric 233 students write thirty pages, on par with UIUC advanced composition requirements.) An additional assumption is that most students will write *more* than these minimums, through lower-stakes assignments and activities, including group work.

What Are Our Long-Term Challenges?

While I have alluded to several challenges throughout my overview of our program, above, I want to close with some more specific—and likely familiar—challenges that UIUC's Rhetoric Program has in store in future years. These challenges I will divide into those specific to our program, and our local conditions, and those that I believe are common to nearly all academic writing programs in the United States.

Challenges at UIUC

First, we are challenged by what our students bring to the table. As previously noted, our rhetoric students are a highly motivated and high-achieving student body (based on their high school learning experiences) who are not all convinced of the value of writing. Like many institutions, we exempt some— but not all—students from the rhetoric course or equivalent Composition I requirement. This can cause rancor amongst students, especially students in the professionally oriented colleges—such as engineering—who would rather "get on" with their major's curriculum and spend less time on general education requirements. I imagine that we are not the only department on campus who suffers from student perception that our courses are unnecessary or repeat work from secondary school curricula. But given that writing instruction is so difficult to measure in the short term—with benefits often popping up months or years after the instruction takes place, as present in both academic and nonacademic settings that draw upon lessons learned in our rhetoric classrooms—we offer a course requirement that certainly can be seen as a nuisance to students rather than a benefit. In past years, under other WPAs, we have had other curricular models and requirements in place for the rhetoric courses, including a common text and standard syllabus and assignments, most recently. Our current retooling, designed around the SLOs and relative instructor autonomy in terms of fulfilling these SLOs, is designed to communicate rigor and specific outcomes of value to a wide range of enrolled students. We are hopeful that this clearer articulation of values and increased emphasis on rigor (for just one example, in our grading standards—which we seek to make more and more transparent with every cohort of new instructors trained) will help students better see both the utility and intellectual weight of the course, as they see their instructors designing course content and readings that reflect their own views on academic writing.

Second, our rhetoric courses are taught by an instructor pool significantly trained in, and most comfortable with, literary study and/or creative writing, alongside a smaller population of instructors enrolled in our thriving writing studies graduate concentration (MA and PhD levels). Again, like many Research 1 universities, we find ourselves training instructors who therefore do not necessarily bring to the classroom a deep knowledge of rhetorical terms or basic rhetorical theory, or have pedagogical interests that lie outside the writing classroom. And to complicate this further, while our curriculum does offer some degree of freedom and choice, we do not allow instructors to teach a literature-based course, nor do we allow them to require creative writing (poetry, fiction, drama) as high-stakes writing assignments. We construct our courses around nonfiction arguments, and as such they are less English (literature) than they are rhetoric—just as the course catalog labels attest.

In addition, with a mix of both MFA and MA/PhD students, we face an instructor pool with varied *amounts* of experience in the classroom; our new MA and MFA students who are granted teaching assistantships go right into the Rhetoric 105 classroom in their first semester, for example, and enroll in the pedagogy course (English 593: Proseminar, described above) alongside more experienced PhDs, who sometimes have taught elsewhere previously. We are not unique here, but we are also fortunate to have a successful and nationally known writing studies concentration, which provides us with MA and PhD students who *do* see their intellectual mission as being tied to the teaching of writing (in some form) and who do often emerge as leaders in our Rhetoric Program's administrative structure or in our Center for Writing Studies program. I would not say that this creates tension, because we have a positive and engaged graduate student community that spans areas of specialization. But it does mean that the WPA always has to have a dual focus in mind: Promote and sustain the Rhetoric Program in such a way that it makes sense for at least three distinct populations of teachers and is also inclusive and mindful of these populations as curricular changes go forward. For us this means creating SLOs with community input (our process was a public one, ultimately involving thirty or more of our instructors who attended monthly meetings to help craft the SLO documents) and creating assessment opportunities that produce public results (for our last assessment in summer 2014, a fifty-plus page document outlining methodology, findings, and recommendations that was distributed to every instructor in our program).

Third, our UIUC Rhetoric Program is part of a giant English department (fifty tenure-track or tenured faculty, over one hundred graduate students, and thirty non-tenure-track faculty) that does not, as a rule, talk together about teaching. This is a condition that we have accepted but continue to brainstorm ways to overcome. As is often the case at institutions as large and centered on graduate teaching as ours, none of our tenure-line faculty teach in our Rhetoric Program. Our English building is so very large and our faculty so many in number that I can go weeks without seeing any particular faculty member unless I seek that person out. We are a very large ship, in other words, and so when we do come together, it is in fragments (tenured faculty deciding promotion cases, for example, or junior and senior faculty socializing within areas of subspecialty, or administrative faculty meeting to discuss department policies and procedures). And usually the gatherings are not specifically focused on pedagogy—even though we weigh teaching considerably in our promotion and tenure decisions as a department and as a campus.

So, this is a problem we in rhetoric have yet to "solve," but it is also not a problem that is ours *alone* to solve. One place where we do see the department potentially coming together more to talk about teaching is through our non-tenure-track faculty group—as these faculty are retained and promoted based primarily *on* teaching, are recently (as of spring 2015) unionized, and are now afforded better rights in their contracts regarding regular professional development. Thus, it is at least my hope that the Rhetoric Program, which employs so many of these faculty, will be able to help the department see its way forward to bridging the conversational gap surrounding teaching through better acknowledgment and inclusion of non-tenure-track faculty in our regular department discussions.

Finally, we are challenged by a campus culture that (often) privileges objective measurements and data-driven research over other types of scholarly production or process-based models. In other words, we value our STEM contributions (in science, technology, engineering, and mathematics) sometimes to a greater extent than we do our humanities work. This means we have departments (units) who would like for us to have specially designated sections of Rhetoric 105 for their own majors, particularly in high-profile and top-ten areas such as engineering. In the past we have sometimes accommodated these requests, with mixed results. But our current Rhetoric Program resists tailoring its courses to meet specific needs in specific units. We instead aim to offer a course that does *not* purport to be all things to all people, as it were. We are cognizant of the research showing that a robust writing curriculum *outside and beyond* the first year is critical to a student's intellectual development and success with writing as a means of learning beyond the rhetoric course. Were we to provide a rhetoric for engineers, for example, and a rhetoric for biophysicists or a rhetoric for geologists—keeping in mind that the last example has never been requested of us—we would suddenly be in the business of creating *many* Rhetoric 105s, with ostensibly different SLOs, specially trained instructors, and varied actual outcomes (and positions in the curriculum). This is not the business we are in. Imperfect as a first-year writing course—anywhere, in any form—arguably is, our ongoing challenge is to keep our curricula holistic and broadly serving, and look to our upper-division writing offerings and writing across the curriculum program to complement and extend what we begin with students in their first year.

Challenges across All U.S. First-Year Writing Programs

That last particular challenge—not to become a servant to many masters, curriculum-wise—is something that many programs face. As I think more broadly, then, about what *all* writing programs likely face in 2015 (the year of this writing), I again find myself gravitating toward three national problems:

1. For those writing programs housed at Research 1 institutions (or any institution wherein the tenure-line faculty are not primarily or also first-year writing teachers), an ongoing challenge is how to balance shrinking or limited graduate enrollments and exploding non-tenure-track positions—some

of which are without benefits—and how these affect staffing, enrollments, and professional development. We all must keep our attention on the ongoing crisis of adjunct labor in our writing programs, especially at institutions where the writing program already runs on the labor of those off the tenure track (whether graduate students, non-tenure-track faculty, or a combination of both, as is the case at UIUC).

2. For those writing programs housed at other types of institutions, there is the ongoing challenge of keeping a vigilant stance regarding the hiring of qualified, trained WPAs who can effectively share best practices in the teaching of first-year writing with faculty who lack knowledge of those practices. While I do not mean to discount the ability of some faculty not trained in rhetoric and composition to lead writing programs on a given campus, we all face the potential denigration of the field of rhetoric and composition through the hiring of WPAs either from other fields or off the tenure track (in terms of job security and programmatic authority among and between campus stakeholders and higher administration). As such, we as an English department believe in the principles set forth in the Portland Resolution, particularly regarding the requisite field knowledge a WPA must have to direct a program.

3. All writing programs must be aware of the phenomenon of dual-credit programs and concurrent enrollment, and these programs' relationship with the Common Core Standards. WPAs must ask, Where does "high school" writing end and "college" writing begin? How does this affect or endanger our university writing programs? WPAs must offer rigorous, localized, engaging first-year writing courses that neither duplicate what may be offered in dual-credit or other secondary school programming nor fail to articulate their own value to students *as* college students, not as individuals who bank credits toward a college degree, off-site and out of the local institutional context.

4. Finally, while it is a significant issue on my own campus, all WPAs must recognize the surge of international students and multilingual learners on campuses nationwide, as well as the trend toward majority-minority enrollments in many writing programs. What are these students' particular needs, and how must our programs evolve to meet them? How does the "standard" curriculum address their needs—if at all? What must we do to make these students feel like a genuine and valued part of our writing programs, and not unwelcome interlopers in them (especially in the case of international students matriculating on campuses previously homogeneous in their enrollments)?

No writing program can champion fixity if it wants to remain healthy and vital both inside and outside its home institution. The snapshot I have provided here will likely change in years to come—either due to new leadership, new conditions for our program's operations, or both. Similarly, no writing program is an island; it is influenced by field scholarship, community discourse, and resource allocations, plus institutional considerations and reconsiderations. What I hope I've been able to sketch out here is the ecology of our UIUC Rhetoric Program that acknowledges room for growth, in ways that might help readers think about the ecologies of their own writing programs—large or small, new or old.

Works Cited

ADE Guidelines for Class Size and Workload for College and University Teachers: A Statement of Policy. Association of Departments of English, 1992, www.ade.mla .org/Resources/Policy-Statements/ADE-Guidelines-for-Class-Size-and -Workload-for-College-and-University-Teachers-of-English-A-Statement -of-Policy.

Addison, Joanne, and Sharon James McGee. "Writing in High School/Writing in College: Research Trends and Future Directions." *College Composition and Communication*, vol. 62, no. 1, 2010, pp. 147–79.

Adler-Kassner, Linda, and Elizabeth A. Wardle. *Naming What We Know: Threshold Concepts of Writing Studies*. Utah State UP, 2015.

Brereton, John. *The Origins of Composition Studies in the American College, 1875–1925*. U of Pittsburgh P, 1996.

Framework for Success in Postsecondary Writing. Council of Writing Program Administrators / National Council of Teachers of English / National Writing Project, 2011.

Glau, Gregory R. "The 'Stretch Program': Arizona State University's New Model of University-Level Basic Writing Instruction." *WPA: Writing Program Administration*, vol. 20, nos. 1–2, 1996, pp. 79–91.

Hansen, Kristine, and Christine R. Farris, editors. *College Credit for Writing in High School: The "Taking Care of" Business*. National Council of Teachers of English, 2010.

Horning, Alice. "The Definitive Article on Class Size." *WPA: Writing Program Administration*, vol. 31, nos. 1–2, 2007, pp. 14–34.

Lamos, Steve. *Interests and Opportunities: Race, Racism, and University Writing Instruction in the Post–Civil Rights Era*. U of Pittsburgh P, 2011.

Lerner, Neal. "Searching for Robert Moore." *Writing Center Journal*, vol. 22, no. 1, 2001, pp. 9–32.

Matsuda, Paul Kei. "The Myth of Linguistic Homogeneity in U.S. College Composition." *College English*, vol. 68, no. 6, 2006, pp. 637–51.

National Council of Teachers of English. "Principles for the Postsecondary Teaching of Writing." Revised Mar. 2015, www.ncte.org/cccc/resources/positions /postsecondarywriting.

———. "Statement on Dual Credit/Concurrent Enrollment in Composition: Policy and Best Practices." Nov. 2012, www.ncte.org/cccc/resources/positions /dualcredit.

———. "Why Class Size Matters Today." Apr. 2014, www.ncte.org/positions/statements/why-class-size-matters.

Portland Resolution. Council of Writing Program Administrators, 1992, wpacouncil.org/positions/portlandres.html.

Rhetoric Instructors Handbook. U of Illinois Undergraduate Rhetoric Program, 2014.

Working Group on Writing Instruction. "The University of Illinois at Urbana-Champaign." *Profiles of Writing Programs in the Alliance for Undergraduate Education*, Alliance for Undergraduate Education, 1993, pp. 16–19.

Yancey, Kathleen Blake, et al. *Writing across Contexts: Transfer, Composition, and Sites of Writing.* Utah State UP, 2014.

4

Breaking Out of the Box: Expanding the WAC Program at Howard University

Teresa M. Redd

Making the case that parents, employers, legislators, and accreditors expect every college graduate to write well, Redd asserts that first-year writing programs alone cannot achieve this goal, nor can writing across the curriculum (WAC) programs that are too often confined to a small set of writing-intensive courses. Like the University of Illinois, Urbana-Champaign (see Ritter's chapter, "Undergraduate Rhetoric at UIUC"), Howard University had to rethink its program after many years of operation. Howard's WAC program succeeded in "breaking out of the box" by recruiting faculty in colleges across campus to hold students accountable for writing well. Redd traces the history of the WAC program at one institution. Then, through a case study of one school at Howard, she presents one way that a WAC program assessed its impact.

Once upon a time there was a historically black university that did not have a writing across the curriculum (WAC) program. However, some professors in the College of Arts and Sciences (COAS)—including the dean—had heard of this thing called "WAC" and wanted to know more. So the dean dispatched an intrepid band of scouts into the wilderness to find out what WAC was all about. The scouts journeyed from Washington, DC, *waaay* up north to upstate New York to visit that remote outpost known as Cornell University. A few days later, the scouts returned with good news: they had seen WAC in action, and they liked what they had seen. So COAS charged the English department with formulating some guidelines, the department presented the guidelines to COAS faculty, and the faculty voted for WAC.

Such were the origins of the WAC program at Howard University (HU). Established by a faculty vote in 1991, the program took hold, but WAC and

HU did not live happily ever after: for years, the program remained "boxed in"—unable to engage most HU faculty and students—until a novel campaign helped the program "break out of the box."

Howard University's WAC Program

The WAC program at HU was built upon the primary tenets of the WAC movement, "writing to learn" (WTL) and "learning to write" (LTW). Although many COAS faculty merely wanted students to learn to write better, the founders of the HU WAC program eagerly embraced WTL, that is, writing as a tool for learning in a discipline. They were persuaded by the theory and research of James Britton and Janet Emig as well as Judith Langer and Arthur Applebee, which holds that writing facilitates critical thinking and thus learning. As Toby Fulwiler and Art Young explain:

> Writing to communicate—or what James Britton calls "transactional writing"—means writing to accomplish something, to inform, instruct, or persuade. Writing to learn is different. We write to ourselves as well as talk with others to objectify our perceptions of reality; the primary function of this "expressive" language is not to communicate, but to order and represent experience to our own understanding. In this sense language provides us with a unique way of knowing and becomes a tool for discovering, for shaping meaning, and for reaching understanding. (x)

As for WAC's other tenet, LTW, the founders of the HU WAC program adopted it but interpreted it in a broader sense than many COAS faculty who equated LTW with mastering grammar, spelling, and mechanics. The founders understood from the work of John Swales, James Porter, Patricia Bizzell, and others that students need to understand a discipline's "discourse community," to which Swales attributes six characteristics:

1. A discourse community has a broadly agreed set of common public goals.
2. A discourse community has mechanisms of intercommunication among its members.
3. A discourse community uses its participating mechanisms primarily to provide information and feedback.
4. A discourse community utilizes and hence possesses one or more genres in the communicative furtherance of its aims.
5. In addition to owning genres, a discourse community has acquired some specific lexis.

6. A discourse community has a threshold level of members with a suit-
able degree of relevant content and discoursal expertise. (24–27)

Unfortunately, having developed such "discoursal expertise" over the years, too
many HU faculty had forgotten what it was like *not* to know these discourse
conventions and thus left students to decipher the conventions on their own.

Given these two tenets, the guidelines for HU's WAC program required
assignments and activities that promoted both WTL and LTW (see the HU
guidelines at http://www.cetla.howard.edu/wac/huprograms.aspx#guide).
Drawing on the "process approach" that was popular in composition pedagogy at the time (Flower and Hayes), the guidelines also required prewriting
and rewriting to achieve WAC's WTL and LTW goals. Therefore, to teach
a WAC-approved course, faculty needed training in pedagogy so that they
could design WTL and LTW assignments and evaluate them effectively and
efficiently. But training was not enough: Faculty who completed training had
to revise a syllabus to demonstrate their mastery of WAC pedagogy and, only
after the syllabus was approved, could they earn certification to teach an official
WAC course. Such qualifications were required at HU since official WAC
courses could fulfill the third writing requirement in COAS—as long as the
class size was limited to twenty to ensure that faculty had time to respond to
frequent writing.

According to the guidelines, WAC courses in COAS had to fulfill six
learning objectives. The writing assignments in a WAC course would enable
students to do the following:

1. Read carefully.
2. Make sense of their lessons.
3. Think critically about the subject matter (especially through analysis
   and synthesis).
4. Organize their thoughts and present them in a comprehensible
   format.
5. Master the textual conventions of the discipline (for example, formats,
   documentation styles, assumptions, rules of evidence).
6. Reinforce the editing skills learned in first-year composition (including proofreading).

To implement these guidelines, the English department appointed an English professor as director (I directed the program starting in fall 1992), and the
director, in turn, appointed an interdisciplinary WAC committee consisting
of faculty representing the humanities, social sciences, and natural sciences

divisions of COAS. However, since the board of trustees had assigned responsibility for the third writing requirement to the English department, the humanities representative also came from the English department. Thus, with the addition of the director of the HU Writing Center, three of the five committee members were English professors. Later, a representative for the health sciences was added, but English faculty still predominated. Then, in 2003, another organizational change occurred: I was appointed director of HU's new teaching and learning center—the Center for Excellence in Teaching, Learning, and Assessment (CETLA). Since I was still directing WAC for COAS, the provost asked me to incorporate the WAC program into CETLA so that more faculty outside COAS could benefit from WAC. I agreed, so WAC became a university-wide program housed at CETLA.

With the move to CETLA, opportunities for WAC faculty development expanded and diversified. Prior to the move, faculty had to wait for an annual three-day summer seminar to participate in WAC training. However, CETLA gradually added a series of three face-to-face workshops that could be taken independently; a hybrid seminar for departments, which began face-to-face and concluded online; and, finally, a fully online seminar that could be offered over three days, a week, or three weeks, one module at a time. Dubbed "WAC Online," this seminar enabled CETLA to conduct last-minute training when necessary and even to invite faculty from the University of South Africa to participate alongside HU faculty one year. According to years of anonymous surveys, the training succeeded in equipping faculty with strategies for designing and evaluating WTL and LTW assignments. In fact, many faculty reported that it had changed not only how they approached student writing but also how they approached teaching in general. One survey respondent summed up the benefits as follows:

> Thank you for such a comprehensive course. I have been introduced to several useful and practical strategies to incorporate writing in my courses. You have also given me new ways to think about how I teach that will help me better develop all my courses and interactions with students.

Compared with training WAC faculty, assessing their classroom implementation and impact proved more challenging. When the first WAC courses were launched in 1993, the WAC committee began collecting student evaluations from WAC courses, evaluations based on the six HU WAC objectives. Periodically, the WAC committee collected faculty evaluations and met with student focus groups as well. However, COAS needed more direct evidence of student progress and proficiency. Initially, the WAC committee conducted a few research studies comparing WAC and non-WAC versions of a course (see http://www.cetla.howard.edu/wac/assessment.aspx#Research_Studies), and a few WAC faculty conducted case studies as well (see http://www.cetla .howard.edu/wac/assessment.aspx#Case_Studies). But the samples were tiny, and the opportunities to compare were rare.

Therefore, in 2001, the WAC committee began recruiting WAC and English faculty to create and then refine a rubric for portfolio assessment, a rubric that grew from a checklist into a detailed descriptive scoring guide (see http:// www.cetla.howard.edu/wac/checklists.aspx). Afterward, using the rubric, pairs of WAC and English faculty members scored portfolios of papers of similar genres and disciplines to assess student progress and proficiency. Although some portfolios were missing or incomplete, both the 2001 and 2008 portfolio assessments suggested that most WAC faculty had been implementing WTL and LTW strategies, especially by assigning journals, summaries, outlines, and multistage research papers and by providing appropriate feedback. Moreover, the portfolio assessments suggested that the faculty were achieving some success in helping students learn to write. For instance, the 2001 assessment of 140 portfolios revealed that the majority of the students had achieved only an "average" level of skill but had progressed in one or more areas of writing (that is, content, arrangement, or style). In contrast, the 2008 assessment of seventy-six portfolios revealed that approximately half of the students had achieved significant progress, but this time nearly a third had achieved a level of "proficient" or "exemplary" while all but two of the remaining portfolios were rated "average."

On the other hand, student evaluations suggested that students were benefiting, above all, from WTL. While students reported that LTW instruction had helped them "organize [their] thoughts and present them in a comprehensible format," an overwhelming majority confirmed the value of WTL assignments, especially as a tool for critical thinking (see http://www.cetla .howard.edu/wac/StudentEvalTables.htm). The following quotations typify the students' responses:

- It really just helped me to reinforce the math. . . . [W]hat I was doing was taking what he said and interpreting it my way so I could understand it when I was writing. (Calculus II)

- This paper helped to solidify the concepts of Ricardo's Corn Theory of Rent. Writing about the subject helped me acknowledge what I did and did not know about his theories and thus helped me greatly to grasp the concepts. For example, the topic proposed for the paper helped me realize the similarities and differences between the different economic theorists: something I probably would not have considered on my own. (History of Economic Thought)

- Yes, because I planned to write this essay I paid closer attention to minute details of the text and the film. If I had not needed to critically analyze the data provided by the film and text, I would have taken everything as entertainment. (Death and Dying)

- When you have to write a paper and have to research a paper and you have to revise, it forces you to internalize certain ideas and concepts. (Black Diaspora II)

- This paper has helped me to fill in some of the missing gaps of Middle Eastern relations. I feel that things are coming together and my understanding of the region is expanding. (Government and Politics of the Middle East)

- Although I learned a lot while in class, I was able to explore this topic more when writing. I looked for important facts we discussed to help strengthen my paper and began to actually connect all the information learned into one puzzle. (Individual and Society)

- As a result of writing this abstract I did read the journal article more carefully than I would have without writing. I think it is a great idea [that] we write an abstract prior to our presentation because it allows us to thoroughly understand what our article is about. It is more so an advantage than a disadvantage. (Parasitology)

- I can truly say that at one point in my journal, I wrote that I finally got it. I was excited. I really did understand. (Ecology Lecture/Lab)

In spite of this evidence, twenty years after COAS faculty voted for WAC, HU's WAC program was in a box, isolated from most HU faculty and students. As of June 30, 2011, only a small number of WAC courses were scheduled each year, averaging only forty annually for the previous ten years. This statistic was not alarming since most WAC faculty did not teach official WAC courses but claimed to use WAC strategies in other courses. What was far more disturbing was the realization that only a handful of HU's approximately 1,300 faculty were engaged in WAC: between 1991 and 2011, only 210 had been fully trained, and only 92 WAC certified. Moreover, the vast majority of those faculty resided in COAS. Without substantial faculty and student engagement in WAC, the status quo prevailed: many students failed to develop professional writing skills or to apply even the basic skills learned in first-year composition to writing assignments in other classes. Too often, I heard students make comments such as "Sure, I outline and proofread for my English professors, but how I write doesn't matter to my biology professor as long as I get the facts straight." Yet that biology professor was likely to be among the many non-English faculty who continually complained, "Why can't these students write? What are you *doing* in the English department?" Writing mattered to these professors, but as long as their students did not *know* it mattered, those students rarely tried to improve or transfer their writing skill. Was there a solution? Perhaps. I called it the "Writing Matters Campaign."

The Writing Matters Campaign

In August 2011, on the twentieth anniversary of HU's WAC program, with the support of the WAC committee and the English department, I launched the Writing Matters Campaign. The campaign sought to improve student writing across the curriculum by recruiting faculty to state high expectations for student writing, to hold students accountable for meeting the standards, and to refer students to resources for writing assistance. These objectives were articulated in a manifesto I called the "Writing Matters Statement":

> Writing is an essential tool for thinking and communicating in virtually every discipline and profession. Therefore, in this course I expect you to produce writing that is not only thoughtful and accurate, but also organized, clear, grammatical, and consistent with the conventions of the field. If your writing does not meet these standards, I may deduct points or ask you to revise. For assistance with your writing, go to the student section of the Writing across the Curriculum (WAC) website.

To join the campaign, faculty had to pledge to include this statement in their syllabi and to post their syllabi in the HU syllabus database for all to see. I also asked faculty to wear a blue and white "Writing Matters" button to class during the first week of the academic year so that students in their classroom and across campus would see that writing mattered to hundreds of HU faculty. In addition, I asked faculty to complete an online Writing Matters survey and to encourage their students to do the same.

Through the campaign, I sought to test a hypothesis: if faculty state high expectations for student writing, students will devote more time and effort to their writing and seek assistance when needed. I was prompted to test this hypothesis by the release of Richard Arum and Josipa Roksa's controversial book *Academically Adrift: Limited Learning on College Campuses*. In their study of more than 2,300 college students who took the College Learning Assessment (CLA) in fall 2005 and spring 2007, Arum and Roksa found that faculty's high expectations were one of the most statistically significant predictors of high CLA scores:

> Even after we have adjusted students' CLA scores for a host of individual and institutional differences, faculty expectations continue to be related to student learning. Students who reported that faculty had high expectations scored twenty-seven points higher on the CLA in 2007 than those who reported that professors had low expectations. (95)

Considering that, on average, students in the study increased their CLA scores by only thirty-four points, this twenty-seven-point gap is, in Arum and Roksa's words, "remarkable" (95). As a result of this finding, Arum and Roksa conclude, "when faculty have high expectations, students learn more"; moreover, they observe, that "having demanding faculty who include reading and writing requirements in their courses . . . is associated with improvement in students' critical thinking, complex reasoning, and writing skills" (93).

Although Arum and Roksa's study suffers from some methodological weaknesses, I could not easily dismiss their findings because there are decades of empirical research documenting the impact of teachers' expectations on student performance. Arum and Roksa cite forty years of sociological research that highlights "how expectations of significant others, including teachers, are important for facilitating students' educational success" (93). Now, there is more recent evidence on a global scale: Cho and Long from Wesleyan University's Quantitative Analysis Center analyzed data from more than two hundred thousand fifteen-year-olds in sixty-four countries from the Programme for International Student Assessment in 2013. They discovered that teachers' low expectations lowered math scores (–65.00 points) and reading scores (–64.56 points). Although they measured the effects of other variables such as socioeconomic status and school environment, they concluded that providing "motivating and high expecting teachers" was as important as "providing a good environment."

John Hattie's research also documents the impact of teachers' expectations. In *Visible Learning*, Hattie synthesized more than eight hundred meta-analyses of influences related to student achievement. Having discovered that the average effect size of the influences he studied was 0.40, he proposed that what works best in education exceeds 0.40. Among the teacher effects, he found that teachers' expectations exceeded this threshold, with an effect size of 0.43, and that "not labeling students" proved even more potent, with an effect size of 0.61. Likewise, he found that giving students challenging but achievable goals produced an effect size of 0.52. Hattie's research also suggests that teachers with high expectations can play a powerful role in student achievement when students' expectations are low. He found that the best predictor of student achievement was the students' expectations in the form of self-reported grades (d = 1.44 or a correlation of 0.80 between students' estimates and their subsequent grades). "If these ratings are too low," he observes, "then such expectations of performance can set limits of what students see as attainable. Hence there is power in teachers setting more challenging goals, engaging students in the learning towards these goals, and giving students the confidence to set and attain their goals" (31). Surely, teachers with high expectations are more likely to motivate students to set and meet high expectations than are teachers with low expectations.

This statement certainly seems to apply to the teachers in the New Zealand primary schools that Rubie-Davies studied. She discovered that, compared with teachers with low expectations, the teachers with high expectations constructed "a framework for student learning by providing children with sufficient and clear explanations and instructions, by carefully orienting students to the task and by linking new concepts to prior learning" (304). Teachers with high expectations also offered more feedback, posed higher-level questions, and created a more positive and caring environment. Thus, she states, "The findings in this study indicate the importance of teachers having high expectations for all their students" (304).

If so, how important are teachers' expectations for African American students? While the impact of teachers' expectations may be small for some groups, in 2005 Lee Jussim and Kent Harber concluded from their review of thirty-five years of empirical research that "powerful self-fulfilling prophecies may selectively occur among students from stigmatized social groups," and they found some of the largest effects on African Americans (131). Likewise, research on language education suggests that teachers' expectations about African American students' language use can be powerful. Citing some of those research findings, Marcia Farr and Harvey Daniels observe that "even the best available teaching methods can fail when implemented by teachers who lack genuine, fundamental appreciation for what students can already do with language" (52). On the other hand, educational researchers have found that teachers who hold *positive* attitudes toward African Americans' language may facilitate the students' acquisition of academic writing skills (Ball; Ball and Lardner; Foster; Ladson-Billings; Moore).

In light of these findings, I hypothesized that students at our predominantly African American institution would try to write better if faculty communicated high expectations about writing. As Hattie's summary of meta-analyses suggests, "goals have a self-energizing effect if they are appropriately challenging for the student as they motivate students to exert effort in line with the difficulty or demands of the goal" (164). Moreover, based on Barbara Walvoord's theory, I expected the quality of students' writing to improve since investing more time and effort could reduce the number of "performance-based" problems in their papers; likewise, seeking assistance from a tutor, peer, or online resource could reduce the number of "knowledge-based" problems (234–40). For instance, a student might take the time to proofread and, consequently, detect and correct performance-based subject-verb agreement errors. On the other hand, another student might take the time to consult with a tutor to understand and practice avoiding knowledge-based subject-verb agreement errors.

With these high hopes, I conducted the Writing Matters Campaign in the fall of 2011, 2012, and 2013. Although some of the survey response rates were too low to be reliable, the results of the fall 2011 and fall 2012 surveys were encouraging.

In fall 2011, 281 faculty from ten schools and colleges took the Writing Matters pledge. Approximately 42 percent (119) of the pledging faculty responded to an anonymous online survey. The results suggest that most respondents implemented the recommended Writing Matters practices, namely, deducting points for deficient writing skills, asking students to revise a deficient draft, and referring students to a tutor or other writing resources. Indeed, faculty reported that they deducted points for deficient writing skills more frequently in their fall 2011 courses than they had the last time they taught those courses—a statistically significant difference (65.6 percent versus 49.5 percent, $p < 0.05$). As for the impact on students, that was difficult to assess since only thirty-six students responded to the anonymous online survey. However, according to the faculty survey, more faculty considered student writing more proficient in their fall 2011 courses than they had when they taught those courses previously—another statistically significant difference (34.3 percent versus 12.1 percent, $p < 0.01$). Moreover, web analytics suggested that more students sought assistance from the WAC website (741 hits versus 354 hits) or the online writing center (106 students versus fewer than 35 the previous year.) Finally, in one social work class where nearly three-quarters of the students responded to the survey, the program appeared to have the desired impact:

According to the fifteen respondents, 93 percent were more likely to compose and revise a rough draft before submitting a paper, 73 percent were more likely to solicit feedback on a draft before turning it in, and 100 percent were more likely to proofread a paper before submitting it when they compared writing for their Writing Matters instructor with writing for instructors who did not tell their students that "they expected thoughtful, accurate, organized, clear, and grammatically correct writing."

In fall 2012, the Writing Matters Campaign yielded similar results. In all, 288 faculty from ten schools and colleges pledged. This time the majority observed that more of the student writing they received was proficient because of the campaign, but only 11 percent of the pledging faculty responded to the faculty survey. The student response was also low: only seventy-two students (taught by twelve instructors) responded. However, 74 percent of the student respondents said, because of the campaign, they were more likely to compose and revise a rough draft before submitting a paper, 70 percent were more likely to solicit feedback on a draft before turning it in, 52 percent were more likely to seek assistance from a writing resource, 89 percent were more likely to proofread a paper before submitting it, and 72 percent were more likely to revise a returned paper.

In fall 2013, once again, the number of pledging faculty (286) barely budged, but I began to notice an increase in participation by school (table 4.1). In the first year of the campaign, I received pledges from 100 percent of the full-time faculty in only one school. The next year, I received pledges from 100 percent in two. Then in the third year, I received pledges from 100 percent of the faculty in three. As table 4.1 reveals, for all three years the School of Social Work achieved 100 percent participation by its full-time faculty. By bringing its part-timers on board, it even doubled its rate of participation during the last two years. Therefore, instead of surveying all of the Writing Matters faculty and students in the fall of 2013, I decided to investigate the impact of the campaign within the School of Social Work. If the campaign was making a difference anywhere, surely it would there.

Table 4.1. Writing Matters Pledges (2011–13)

Campaign Year	*No. of Pledges*	*100 Percent Pledging Schools*
2011	281	Social Work
2012	288	Social Work, Nursing
2013	286	Social Work, Business, Law

Indeed, the School of Social Work had a story to tell. Nearly half (44 percent) of the school's forty-three full-time and part-time faculty responded anonymously to a fall 2013 survey that asked them to consider what had happened over the years since they joined the Writing Matters Campaign. (Appendix A reproduces the faculty survey questions.) According to the survey results, at least three-quarters said they had included the Writing Matters Statement in their syllabi, announced the campaign in class, and sported blue and white Writing Matters buttons during the first week of class (table 4.2). Nearly all assigned, on average, at least two papers per course, almost two-thirds assigned three papers, and approximately a third assigned four or more papers. True to their pledge, nearly all of the faculty members had deducted points for deficient writing more often than they had prior to the campaign. In addition, approximately half had asked students to revise more often, half had referred students to writing resources more often, and a third had referred students to a tutor more often. Moreover, although only a few had attended a WAC workshop, a little more than half of the faculty respondents had visited the WAC website, and nearly two-thirds had consulted other writing resources after joining the campaign.

So did the campaign make a difference in the School of Social Work? Did the students change their composing processes? Did they seek assistance? Yes, if we accept the results of the anonymous student survey, to which 65 percent of the 198 social work students responded in fall 2013. (Appendix B reproduces the student survey questions.) Nearly 88 percent of the respondents

Table 4.2. Social Work Faculty Implementation (2011–2014) in Rank Order (n = 19)

Writing Matters Behavior	*Percentage*
Included Writing Matters Statement in syllabus	74
Announced the campaign in class	84
Wore Writing Matters button first week	84
Deducted points for deficient papers	84
Consulted writing resources (other than WAC website)	63
Referred students to writing resources	58
Visited the WAC website	58
Asked students to revise more often	53
Referred students to a tutor	32
Attended a WAC workshop	11

confirmed that they were aware of the campaign because of the faculty's syllabi, announcements, and/or buttons, so this analysis focuses on their responses. Approximately 3 percent of those who were aware of the campaign said they had been taking social work courses since 2010, 12 percent since 2011, 36 percent since 2012, and 39 percent in 2013. Therefore, nearly half had been enrolled in the School of Social Work for at least one and a half years. Nearly three-quarters (74 percent) described their instructors' expectations for student writing as "high."

On the surveys, the students attributed the following changes in their writing behavior to the campaign:

- More than half said they were more likely to prepare a rough draft.
- More than half said they were more likely to solicit feedback on a draft.
- More than two-thirds said they were more likely to proofread.
- Nearly a third said they were more likely to seek assistance.
- At least half said they had spent more time preparing their papers.

One student aptly summed up the situation by saying, "Writing matters no matter what courses you are taking."

Given such testimony from the students, it is not surprising that nearly three-quarters (74 percent) of the faculty respondents concluded that the quality of student writing had improved since they joined the campaign, and none said it had not. Furthermore, 79 percent of these faculty members rated their students' writing as "proficient," and 7 percent called it "exemplary." As one professor declared on CETLA's anonymous survey, "It's working!"

Institutional Change

Because of these findings, the WAC committee agreed to focus the fall 2014 campaign on pledges by units instead of individuals. Surely, the power of teachers' high expectations would increase when those teachers were working together toward a common goal and when students had to meet Writing Matters standards in nearly every class they took. So in fall 2014 we asked deans and chairpersons to initiate discussions of the Writing Matters Campaign at their August faculty retreats and meetings. If a majority of the faculty agreed, the dean, program director, or chairperson would sign and email me a pledge for the unit instead of collecting individual pledges. (I had to exclude the College of Medicine because it began classes too early to participate, and the

Graduate School because it had no faculty of its own.) In all, I received unit pledges from seven of the thirteen eligible schools. Consequently, participation in the Writing Matters Campaign increased from 286 in fall 2013 to 413 in fall 2014. The following fall, nine schools pledged, increasing participation to a record-breaking 908 (table 4.3).

More important, we finally began seeing more institutional changes:

- The School of Business and the School of Law added the Writing Matters Statement to the templates for their syllabi.
- The School of Divinity launched the course Introduction to Ministry and Theological Writing.
- The School of Social Work set up a writing workshop for all incoming students.
- The Graduate Nursing Program began requiring incoming students to enroll in the Graduate School's writing course their *first* semester if they did not pass the Graduate School's writing test. Moreover, the graduate nursing faculty raised the passing test score and made passing the test or course a prerequisite for the required research methods course.
- The Office of the Provost funded more tutors for the HU Writing Center to meet the growing demand for its services.

Table 4.3. Writing Matters Pledges (2011–2015)

Campaign Year	*No. of Pledges*	*100 Percent Pledging Schools*
2011	281	Social Work
2012	288	Social Work, Nursing
2013	286	Social Work, Business, Law
2014	413	Social Work, Business, Nursing, Law, Communications, Divinity, and Dentistry plus departments in Arts and in Sciences and in Education
2015	908	Social Work, Business, Nursing, Law, Communications, Divinity, and Dentistry, Arts and Sciences, and Pharmacy plus departments in Education and in Allied Health

Granted, the campaign has not achieved all of the WAC program's goals: it has not substantially increased faculty participation in WAC training, addressed WTL, or elicited 100 percent participation from all eligible schools. Nevertheless, the Writing Matters Campaign is engaging faculty in the slow process of building an institutional culture of writing excellence. So we are, gradually, "breaking out of the box."

Appendix A: Social Work Faculty Survey

Communication

1. During the past three years, did you do the following in one or more of your courses? (Click ALL that apply: (a) wore a Writing Matters button, (b) announced Writing Matters, (c) included the Writing Matters Statement in the syllabus, (d) none of the above)
2. How would you compare the way you communicated your expectations for student writing during the past three years with the way you communicated your expectations before joining the Writing Matters Campaign? (Choices: the same, clearer, much clearer, N/A)

Implementation

3. Did you implement any of these practices MORE often during the past three years than you had prior to joining the Writing Matters Campaign? (Click ALL that apply: (a) deducted points for deficient writing skills, (b) asked students to revise a deficient draft, (c) referred students to a tutor)
4. On average, how many papers did you assign in EACH Writing Matters course during the fall 2013 term? (Choices: 0, 1, 2, 3, 4 or more)
5. Since joining the Writing Matters Campaign, have you taken any of the following steps in order to fulfill the Writing Matters goals? (Click ALL that apply: (a) visited the WAC website, (b) attended a WAC workshop, (c) consulted other writing resources, (d) none of the above)

Impact

6. On average, do you think the quality of student writing in your Writing Matters courses has improved since you joined the Writing Matters Campaign? (Choices: yes, no, I don't know)
7. If so, on average, how would you rate the quality of student writing in your Writing Matters courses? (Choices: exemplary, proficient, average, minimal, deficient)
8. Please type any additional comments below.

Appendix B: Social Work Student Survey

1. I have been taking courses in the School of Social Work since______?
   (Choices: 2010, 2011, 2012, 2013)

Communication

2. Since the fall of 2011, approximately how many of your Social Work instructors have worn a Writing Matters button in class? (Choices: 0, 1–2, 3–4, 5–6, 7–8, 9–10, 11 or more)
3. Since the fall of 2011, how many of your Social Work instructors have announced the Writing Matters policy in class? (Choices: 0, 1–2, 3–4, 5–6, 7–8, 9–10, 11 or more)
4. Since the fall of 2011, how many times have you seen the Writing Matters Statement or a similar passage in a syllabus for one of your Social Work courses? (Choices: 0, 1–2, 3–4, 5–6, 7–8, 9–10, 11 or more)

Implementation

5. How would you describe most of your Social Work instructors' expectations for student writing since the fall of 2011? (Choices: low, medium, high)
6. Since the fall of 2011, how many of your Social Work instructors have deducted points on your papers because of deficient organization, style, grammar, spelling, or mechanics? (Choices: 0, 1–2, 3–4, 5–6, 7–8, 9–10, 11 or more)

Impact

7. How has writing for your Social Work instructors compared to writing for instructors who did not state high expectations for student writing? In other words, since the fall of 2011, for your Social Work instructors, were you (less likely, as likely, more likely, N/A) to do the following? a) solicit feedback on a draft from the instructor, classmate, friend, or family before turning it in, (b) proofread a paper before submitting it, (c) compose and revise a rough draft before submitting a paper, (d) revise and resubmit a returned paper, (e) seek assistance from the Writing across the Curriculum (WAC) website, Writing Center, or another writing resource
8. Have you spent more time preparing papers for your Social Work instructors than you expected to before you were aware of the Writing Matters Campaign? (Choices: yes, no, N/A)

9. How many times have you accessed the Writing across the Curriculum (WAC) website this semester? (Choices: 0, 1, 2, 3, 4 or more)

Conclusion

10. How important do you think writing is in your future? (Choices: not important, somewhat important, important, very important)

Works Cited

Arum, Richard, and Josipa Roksa. *Academically Adrift: Limited Learning on College Campuses*. U of Chicago P, 2011.

Ball, Arnetha F. "Community-Based Learning in Urban Settings as a Model for Educational Reform." *Applied Behavioral Science Review*, vol. 3, 1998, pp. 127–46.

Ball, Arnetha, and Ted Lardner. "Dispositions toward Language: Teacher Constructs of Knowledge and the Ann Arbor Black English Case." *College Composition and Communication*, vol. 48, no. 4, 1997, pp. 469–85.

Britton, James. "Writing to Learn and Learning to Write." *Prospect and Retrospect: Selected Essays of James Britton*, edited by Gordon M. Pradl, Boynton/Cook, 1982, pp. 94 111. Reprinted from *The Humanity of English: NCTE Distinguished Lectures*, National Council of Teachers of English, 1972. Web.

Cho, KangWon, and Daniel Long. "The Effect of Teachers' Expectations on Students' Academic Achievements from a Global Perspective." Quantitative Analysis Center, Wesleyan University, 1 Jan. 2013. Web.

Emig, Janet. "Writing as a Mode of Learning." *College Composition and Communication*, vol. 28, no. 2, 1977, pp. 122–28.

Farr, Marcia, and Harvey Daniels. *Language Diversity and Writing Instruction*. National Council of Teachers of English, 1986.

Flower, Linda, and John Hayes. "A Cognitive Process Theory of Writing." *College Composition and Communication*, vol. 32, no. 4, 1981, pp. 365–87.

Foster, Michele. "Effective Black Teachers: A Literature Review." *Teaching Diverse Populations: Formulating a Knowledge Base*, edited by Etta R. Hollins et al., State U of New York P, 1994, pp. 225–41.

Fulwiler, Toby, and Art Young. *Language Connections: Writing and Reading across the Curriculum*. National Council of Teachers of English, 1982.

Hattie, John. *Visible Learning: A Synthesis of Over 800 Meta-analyses Relating to Achievement*. Routledge, 2013.

Jussim, Lee, and Kent D. Harber. "Teacher Expectations and Self-Fulfilling Prophecies: Knowns and Unknowns, Resolved and Unresolved Controversies." *Personality and Social Psychology Review*, vol. 9, no. 2, 2005, pp. 131–55.

Ladson-Billings, Gloria. "Who Will Teach Our Children? Preparing Teachers to Successfully Teach African American Students." *Teaching Diverse Populations: Formulating a Knowledge Base*, edited by Etta R. Hollins et al., State U of New York P, 1994, pp. 129–42.

Langer, Judith A., and Arthur N. Applebee. *How Writing Shapes Thinking: A Study of Teaching and Learning*. NCTE Research Report no. 22. National Council of Teachers of English, 1987.

Moore, Renee. "Teaching Standard English to African American Students: Conceptualizing the Research Project." *Bread Loaf Rural Teacher Network Magazine*, Summer 1998, pp. 12–15.

Rubie-Davies, Christine M. "Classroom Interactions: Exploring the Practices of High- and Low-Expectation Teachers." *British Journal of Educational Psychology*, vol. 77, 2007, pp. 289–306.

Swales, John M. "The Concept of Discourse Community." *Genre Analysis: English in Academic and Research Settings*. Cambridge UP, 1990, pp. 20–33.

Walvoord, Barbara E. Fassler. *Helping Students Write Well: A Guide for Teachers in All Disciplines*. 2nd ed., Modern Language Association of America, 1986.

5

"Who Are the Specialists?" The Dependent Independent Writing Program at Cornell

Paul Sawyer

Describing the Knight Institute for Writing in the Disciplines at Cornell University, where the director is always a professor from a separate home department who serves for a three-year term, renewable, is a complex task. It is a "dependent independent program"—in that it stands alone from any department on campus yet depends on multiple collaborations with those departments and other entities across the university. The writing experts at Cornell—the people who have sustained a nationally prominent and innovative program for the past three decades and more—are colleagues who are neither tenured nor professors. Several have even written highly regarded books on the lived experience of the writing classroom. Sawyer describes the strengths of the Cornell program while acknowledging challenges that have emerged.

My colleague Keith Hjortshoj has described the Knight Institute for Writing in the Disciplines as a "dependent independent program"—independent from any single department yet dependent on multiple collaborations with departments and other entities across the university. For that reason, we're in a position to generate new collaborations by an almost natural process. (In this chapter I'll be using "WAC," or writing across the curriculum, to stand for "WID," or writing in the disciplines, the acronym our name would require.) I'll argue that this process gives the Knight Institute staff a function and an expertise that differ from the usual idea of a specialist in writing, particularly that espoused by compositionists. I conclude with a polemical section that looks broadly at the relationship of writing pedagogy to the deterioration of academic working conditions nationwide.

Ezra Cornell's University

No single reason explains why Cornell University, by the early 1980s, spontaneously developed one of the first WAC programs in the country. But the historic heterogeneity of the Ithaca campus was surely a factor. The purpose of the Morrill Act of 1862 dovetailed with Ezra Cornell's stated intention to "found an institution where any person can receive instruction in any subject."[1] The earliest version of the campus included a working farm on the present site of the central research library; Cornell also required that each undergraduate practice a manual craft. In 1870 the definition of "any person" expanded to include women. Today the Ithaca campus contains seven undergraduate colleges that include an endowed College of Arts and Sciences; an endowed College of Engineering; an endowed College of Art, Architecture, and Planning; and four state colleges: Human Ecology, Industrial and Labor Relations, Hotel Administration, and Agriculture and Life Sciences. In fulfillment of the requirements of the Morrill Act, the agricultural school administers the Cornell Cooperative Extension, which has an office in every county in New York State—our major response to the requirements of the Land-Grant College Act. When the phrase "Ivy League" was coined in the 1930s, Cornell was the only land-grant institution among the eight so nicknamed.

Ezra Cornell's statement of intent still stands as his university's virtual motto, marking the hopeful union, in his view, of the practical and the theoretical, and of democracy and higher learning. His plan was not uncontroversial. After a visit to the United States, Matthew Arnold wrote in the preface of his canonical work *Culture and Anarchy*, "The university of Mr. Cornell, a really noble monument of his munificence, yet seems to rest on a misconception of what culture truly is, and to be calculated to produce miners, or engineers, or architects, not sweetness and light." (22). The comment isn't much remembered today, but early Cornellians took pride in mocking its pretentiousness. Nevertheless, the university's composite nature generated tensions in status between the endowed colleges and the state colleges, which partly coincided with the difference in tuition (lower for in-state students applying to the land-grant portion of the campus) and derived ultimately from the old distinction between theoretical knowledge and the applied fields—the distinction implied in Matthew Arnold's jeering comment. The university that contained a Department of Poultry Science and granted tenure for research into the hospitality industry also needed the intellectual capital to compete with Harvard, Princeton, and Yale for its "best" students. Cornell's international partnerships now include a medical college in Qatar (whose writing program staff collaborate with us) and a research center in Manhattan jointly administered by Technion, an Israeli engineering firm with ties to the Israeli ministry of defense—developments that make the notion of "mechanical arts" seems positively quaint. In this way, Ezra Cornell's goal of "instruction in any subject" has ended up embodying all the contradictions in American professional education.

Today, each college exists under a central umbrella yet maintains separate requirements and procedures—much like the fifty American states in relationship to Washington. This institutional version of *e pluribus unum* is the pride of Cornell when viewed on one of its sides (autonomy) and also the bane of Cornell when viewed on another (bureaucratic confusion); the bane includes obstacles to giving students credit in other colleges or to dividing budgets or to forming interdisciplinary partnerships. As a result, the university's intellectual variety can lie undiscovered. A student studying painting and sculpture in the architecture school may never run into the student who's learning operations engineering across campus, or the student learning crop sciences in the Ag Quad, or even know there's an equine research lab or a lab of ornithology. Often the left hand is doing in one corner what the right hand is trying to get funding to do in another.

It's often said that the only academic experiences shared by all Cornellians are the swim test and the first-year writing seminar (FWS). (Though we serve all undergraduates, the Knight Institute is an entity of the College of Arts and Sciences; our chief funding comes from the college and from three crucial grants from the James S. and John L. Knight Foundation.) And so the FWS curriculum must make a variety of courses available—a lot of them—to satisfy future engineers, potential hotel managers, and the seven hundred undergraduate business majors, as well as "Artsies." Our participating departments, in addition to the usual suspects, include linguistics, government, philosophy, anthropology, music, history, and area studies. (Even mathematics has offered a seminar; the graduate instructor went on to edit a brilliant series of annual volumes on the best essays on mathematics.) A graduate instructor in English is likely to teach three or four seminars before encountering a prospective major in her field. In spring 2015, we awarded the Buttrick-Crippen Fellowship—given to the graduate student who submitted the best new course proposal—to a biomedical engineer who will teach the first writing seminar in the history of his college (the other prizes awarded went to an architectural historian and an astronomer).

One Writing Program

A WAC program like ours is based on the assumption that writing in the disciplines is best taught by people who themselves write in those disciplines.[2] The assumption is plausible, but as many other institutions have found, the practical obstacles can be formidable. Departments can lose interest in the arduous task of recruiting and training graduate instructors specifically to teach writing. Faculty buy-in seems particularly chancy when it's so tempting to let graduate instructors do this labor-intensive work. The success of the WAC model at Cornell rests on two structural features—the system of matching teaching assistant (TA) stipends and the instructors' training courses—but mostly on a third, less tangible feature: the quality of working conditions for the staff.

First-Year Writing Seminar Partnerships

A system of incentives and trade-offs allows the Knight Institute to tie a participating department's teaching of FWSs to financial support for that department's graduate students. TA stipends come to us from the College of Arts and Sciences; participating departments have stipends of their own, but they can't support all their graduate students without our contribution. TA stipends are therefore the coin of our realm. One condition of these matches is a pledge by departments to offer a number of faculty-taught seminars; as a result, of the approximately 330 FWSs offered each year, about one-third are taught by faculty (lecturers and professors). By this means, we ensure a substantial level of faculty participation in the teaching of writing; we allow students to select seminars according to their interest in a wide variety of topics; and we help support about 150 graduate students each year. At the same time, the range of participating disciplines is restricted because of the way graduate students in the sciences are funded. Young scientists usually work as lab researchers or as TAs in large lecture courses—a form of teaching markedly less labor intensive than the teaching of writing. English graduate students and faculty teach 40 percent of FWSs; the percentage taught by literature departments is well over the majority. This feature, combined with the fact that the directors have all been male faculty members in English or comparative literature, means that the program's structure is traditional in many ways—which is one reason why the Writing in the Majors program reached out to the sciences (as I discuss below). A greater shortcoming is the enrollment cap for FWSs. Because a limited number of departments teach students from seven colleges, our classes are large: the cap of seventeen students per section (and for many now, eighteen) exceeds the cap on first-year seminars in our peer institutions, in some cases by a long shot.

The Writing Workshop

The importance of the Writing Workshop, both to our program and to the Cornell community, is immense. For nearly three decades, Joe Martin has served and advocated for Cornell's most vulnerable students—the insecure, the culturally or linguistically marginalized, the ones with an accent or a learning disability—usually from behind the scenes. The offices of Writing Workshop staff, which are grouped around a central reception area, are places where a variety of small conversations take place—a group of students from Korea and other countries who are taking a developmental writing seminar, a graduate student with a difficult research project, a Cornell employee working on a cover letter, a student preparing a PowerPoint presentation on her three years' work as a Public Service Scholar. All of Cornell's writing support for international undergraduates takes place here, and nearly all its tutoring and consulting. The meeting places of the Writing Walk-In Service are dispersed across campus at strategic locations. On a given day, Tracy Carrick, director of the service, may be holding a training session with undergraduate tutors, fixing glitches in the scheduling program, or calming an anxious student in her office. Her work in training tutors has been a key element in our developing new collaborations.

Training the Instructors

I've never been ashamed of the fact that Cornell freshmen are sometimes taught by graduate students—because grads receive first-rate training. Unless they have substantial teaching experience, all first-time graduate instructors must take a six-week course in writing pedagogy, either in the summer or fall. The theory of WAC pedagogy is sharply tested here, in a training course that provides a common syllabus and format for graduates coming from various disciplines. Writing 7100 has evolved over the years and is now consistently popular, but it's only the first step in the professional development of aspiring college teachers.[3] In departments like English, grads design their own courses after a year's teaching; in all departments, they receive regular faculty mentoring, compete for teaching prizes, and accumulate solid teaching credentials for the job market. Writing 7100 has already trained more than a generation of teachers; it's work that spreads beyond the Cornell campus and into years of teaching, as our students in turn influence their colleagues and their students and so on in an indefinite expansion. For that reason, Cornell's training course is the single feature of our program that I would most recommend to similar institutions. (It's worth pointing out that in a severe academic buyers' market, the thorough training, along with the reputation of Cornell's program, continues to make *some* difference in our instructors' chances at a job.)

Staff

Lecturers in the College of Arts and Sciences are hired for a three-year contract, renewable once upon a review, after which they can be promoted to senior lecturer, with five-year renewable contracts thereafter (always following a review). We've not yet lost staff because of budgetary crises or other changes in policy, nor have we failed to renew staff because of poor reviews. "Tenure without tenure" describes the position accurately, which in reality amounts to substantial job security—though obviously lecturers remain vulnerable in times of contraction in ways that tenured faculty do not. Full-time teaching for lecturers at Cornell is the *equivalent* of six courses per year; full-time Knight Institute staff teach four sections or fewer of the Writing Workshop's developmental writing seminars (Writing 1370 and 1380), along with either of the teacher training courses or an occasional writing course outside the Writing Workshop. Nearly all staff members have a specific expertise or administrative responsibility. We do no casual hiring; the exception is freshly minted PhDs who are without a job and who are eligible to serve one year—the limit for any academic employee hired without a search.

The Knight Institute's director, however, must be a professor from a home department, since the program's designers wanted to ensure that teaching writing would be taken seriously in an institution governed by academic status (which of course depends on the distinction between research positions and teaching positions). No one at that time considered hiring a director with a PhD in the then-new field of rhetoric and composition, which would have raised the sticky question of how research in this field would be evaluated (read: taken seriously) by the literature faculty in his likely home department who would be voting to tenure him. (For the record, in addition to a retired colleague with a PhD in anthropology, Knight Institute employees include a PhD in English from Princeton, a rhetoric and composition PhD from Syracuse, a history PhD from Chicago, a TESOL [teaching English to speakers of other languages] degree holder, and a former English graduate student without a degree, among others.) The rank of lecturer means that the staff, in my opinion, are underpaid for the work they do relative to professorial positions; on the other hand, the lecturer staff have the respect of professorial faculty because they're specialists in writing, even though there's a sense in which none of us are "specialists"—as I describe next.

Writing Nonspecialists

Every writing program director has been accosted at some point by a professor who has discovered that his honors students are "terrible writers" and blames freshman writing for failing in its duty. Why, the complaint goes, didn't you take care of this problem for once and for all in the first year? But when pressed to elaborate on his students' lacks, the colleague tends to become more vehement rather than more specific: the students just "can't write" or, even, "can't write a simple sentence." The first assumption in the question, of course, is the disproven view that writing is a basic and universal skill, which, when taught, lasts through any and all levels of demand. On the contrary, we know that the top performers in a FWS may collapse under the different challenges of a senior thesis. The second assumption is that "good writing" is a universal, obvious "something," whereas in fact it's a portmanteau concept. The manifest meaning is, often, basic grammatical competence—the rules of grammar and usage one would learn, say, in a foreign language course if the language were English. Like infotainment media, in their perennial discoveries of a crisis in writing, the frustrated professor isn't really referring to grammatical competence, but nevertheless his complaint seems to invoke that category, as a default position. A second meaning of the phrase "good writing" relates to a undefined scale of aesthetic excellence—*style*, in a word—which can be praised by terms like eloquence, rhetorical mastery, clarity, conciseness, and the like. Good writing in those two senses (correct grammar, English-y eloquence) is not something my colleague does; his time is too valuable to teach the basics at this level, and he's not trained to teach style. Neither of these is the way a writing professional defines "good writing." We might think of it as a student's successful attempt to "invent the university," in David Bartholomae's phrase—to assume the authority of the academy by internalizing the codes of its discourse, or as I'd prefer to put it, the codes of various overlapping academic discourses. For this last, best sense of "good writing," a WAC program has no experts; the frustrated academic—he and no one else—will have to teach his seniors himself. This is the point my colleague Keith Hjortshoj brilliantly establishes in his recent essay "An Alternative History of an Interdependent Writing Program."

Keith is the founder of Cornell's program called Writing in the Majors (WIM). He has told me that in his first-year English course, he received a C; he never took another English course, instead earning a PhD in anthropology. Years later, as a teacher of writing in the Knight Institute, he was asked by my predecessor Harry Shaw to design a program for upper-division writing courses. He decided to take an anthropologist's approach. Instead of explaining to faculty in various fields what good writing is according to a professor of English, he visited departments to listen and gain a sense of their academic culture: how they wanted to use the TA stipends he could provide them and what *they* thought good writing looked like. From those conversations came a new set of writing-intensive courses, with several varieties in format. Incidentally, Keith insisted from the start that WIM courses *not* satisfy a formal writing requirement of a college. The stipulation encourages teachers to integrate writing into the content of the course rather than attaching it as a separate skill to be mastered, an additional hoop.

In his essay Keith describes meetings with two WIM professors in his office, one a historian and one a physicist, who had separate complaints about their students. Both faculty members assumed their students had picked up the respective—but almost contradictory—bad habits from their first-year writing class. Keith pointed out that since their own fields "created and maintain" the "rules" for writing and thinking in their disciplines, the faculty themselves were responsible for teaching them to students:

> There is no God of Writing, no central authority over written language responsible for creating and solving the problems that all of us encounter. Nor, then, are there legitimate priests of writing, in the English Department, in the "independent" writing program, or elsewhere. The language we use belongs to all of us. The ways we use this language and expect our students to use it are our responsibility, and if we don't teach our students how to meet our expectations, we can't expect anyone else to do so (Hjortshoj, "Alternative History" 79).

Faculty are competent to teach writing in the disciplines because they are themselves writers. No one denies the last part of this sentence; chemists and engineers, for example, have told me they spend over half their professional time writing. But at the same time, faculty don't always believe they were *taught* to write in their fields. They seem to have learned their disciplinary language by osmosis (the metaphor is now ubiquitous) without ever describing what they now recognize "instinctively" as good writing in their field.

This observation is one answer to the question, If writing staff do not represent the god of good writing, what do they do? Instead of legislating the conditions of good writing, they can help faculty come to self-awareness about the conventions and qualities of writing in their own fields. For example, my colleague David Faulkner, who now directs the FWS program, sets up assessment meetings with faculty in participating departments—meetings in which a discussion of grading often turns into a discussion about the department's strategies for teaching the "rules" of their own discourse to their majors. This kind of group work extends to insights about professors' own writing as well. Given a more flexible idea of the varieties of possible writing, some faculty turn out to be better writers than they realize—as I learned several summers ago in the Faculty Seminar in Writing Instruction.

Until the advent of the Center for Teaching Excellence, our Faculty Seminar was for several years the only forum at Cornell for faculty to have extended conversations about pedagogy with their peers. The premise of the seminar, like that of Writing 7100, is counterintuitive: that faculty colleagues, when asked to present a detailed assignment for a prospective writing course in, say, city and regional planning or in plant science, can both learn from the problems facing, say, a teacher of city and regional planning or plant science, and give helpful advice. In fact, the mixture of disciplinary perspectives is often the most popular feature of a course that draws high praise in evaluations. For his presentation, a physics professor decided to take the problem he assigns every Physics 101 class—a basic question about the angle at which one must throw a projectile in order for it to land in a given place—and convert it into a writing assignment that would require students to explain each step. The instructor did the same himself and, to his surprise, discovered something new about this venerable problem (he found out that gravity is irrelevant to the solution). This was the main insight. Another came during the discussion, when the professor criticized his own style as flat and inelegant. Since I had personally found the writing lucid and energetic, I asked him if he and his colleagues ever spoke this way when working on problems together in the lab. It turned out that he had indeed imitated the style of physicists in conversation about their work. We were then able to talk about the markers of scientific conversation, its mix of colloquial and technical, its shortcuts, its ways of conveying meaning quickly, and so forth. The benefits of an incident like this do not run in a single direction, from the Knight Institute outward. From the sciences in particular, I've gained a sharper sense of the limitations of the older model of education that posits learning as a private contract between a student and teacher, an owned credential, an exercise isolated from the social context of "real-world" research and discovery. I've learned from scientists who teach students to collaborate in groups, to present or share their learning in speeches or posters, and to engage in various ways with audiences and outcomes. This model of learning is far from the notion of writing instruction common in composition studies, which roots itself in the experience of the first year of college.

In introducing new instructors to our training program, I sometimes define WAC by using a version of the older composition model as a straw man: a single, gigantic course, offered through the English department and divided into scores and even hundreds of untitled sections. I suggest that "freshman comp," when separated from the rest of the university curricula, tends to be inherently unequal: writing is easily treated as a remedial skill, like a swim test, its administration housed in the inferior annex of an English department and its instructors given pay, status, and rank inferior to "real" faculty. I'm aware that this crude contrast ignores the problem of skill levels among students.

For a student body poorly prepared for writing in college, a program will need more writing-intensive courses that keep content to a minimum—like our own Writing Workshop seminars, one of which I'm teaching with genuine pleasure at the time I write this. The greatest problem with the freshman comp model, in my view, is not its ignoring of disciplinary contexts but its tendency to foster adjunct labor (which I discuss in a later section). However, I also worry about a conceptual problem: a too-rigid conception of freshman comp that leads to a false distinction between writing and reading.

To my surprise, I recently learned that the truism "good readers make good writers" implicitly goes against an implicit assumption of composition studies, as that field is now constructed. To see this, one need go no farther than the field's semiofficial "Principles for the Postsecondary Teaching of Writing," revised in 2013 by the Executive Committee of the Conference on College Composition and Composition (CCCC). Using strained syntax, the 2013 version of the document spells out eight principles of sound writing instruction without once using the words *reading*—or, even more surprisingly, *research*.[4] A similar restriction in research in the field suggests to me that compositionists are—understandably—basing their hopes of job security on a definition of expertise that can be linked to an English department. But the sacrifice in breadth is unnecessary—and does little, in the present economy, to preserve jobs and salaries. In a welcome recent study, *Securing a Place for Reading in Composition*, Ellen Carillo argues for bringing reading back into the composition classroom. Carillo makes her argument by discovering a "golden age" of writing about reading by compositionists of a previous generation—without, however, mentioning the simultaneous growth of WAC programs. Like Carillo, I believe the rhetoric and composition degree should also be a degree in teaching reading—but not (I would argue) the reading of poems and fiction only (which would mean tying writing pedagogy even more tightly to the apron strings of English) but potentially of any and all forms of language that students are likely to encounter. Compositionists, like the rest of us, should dive into the world of language, teaching anything from Facebook to op-eds, from conversations among physicists to patterns of political propaganda, from the implicit messages of photographs to the tropes of classical rhetoric. Since interpreting the reading codes of the public sphere is a survival skill for any of us and particularly for our students, these readings could enliven the composition classroom and give it relevance.

A final benefit of the WAC model is institutional. Because Knight Institute programs are already collaborative, new collaborations have built easily on the old ones, often at the initiative of people or units that have worked with the writing staff. These partnerships would not be imaginable if our program had remained a subset of an English department.

From Writing across the Disciplines to Collaborations across Units

The five collaborations I describe below respond to teaching goals we couldn't have reached by ourselves: to deepen the teaching of undergraduate research, to reach more traditionally underserved students, to develop pedagogy for both native speakers and international students in the Graduate School, to forge pedagogical connections beyond the liberal arts, and to use writing as a form of active engagement with communities beyond the campus. In several cases, the obstacles have been more obvious than the successes.

Research Librarians

As everyone recognizes, despite the extraordinary intellectual advances made possible by the internet, doing research is not the same as owning a personal computer. A student in college working on her tenth research paper may still be doing what she did in middle school—compiling online sources and dropping citations into her text without having learned to evaluate sources, to engage them, to pose a research question, or most importantly, to formulate a purpose. An ongoing study by Sandra Jamieson in the Citation Project, which details the features of perfunctory college research papers, also suggests that the research skills of incoming freshmen are unlikely to improve if instructors give perfunctory assignments. The extraordinary sales of Graff and Birkenstein's handbook *They Say, I Say* is one obvious response to this felt need, particularly in humanities and conceptual social science majors.

When I became director of the Knight Institute, I knew we couldn't on our own develop a systematic plan to teach research at all levels. But we made a start by collaborating directly with Cornell's research librarians. One of my colleagues had read about the use of "embedded librarians" in the writing classroom. When we approached the research librarians, several told us they considered collaborative learning to be the wave of the future in their profession. Many had been dissatisfied with the constraints of the single-session library orientation, which is usually limited to tips about the use of basic search engines. Obviously, writing a research essay draws on two separate competences, not one: knowing where to find the best sources (information literacy, which is traditionally the expertise of university librarians) and knowing what to do with those sources (traditionally the expertise of faculty). It was less obvious to me that when we assign research essays unthinkingly, we generate a false distinction: between the "real" work we teach—analyzing, synthesizing, conceptualizing, composing—and the mechanical work of locating stuff. But both processes are creative, part of the indissoluble activity of intellectual discovery.

The FWS called Writing and Researching in the University has been taught in succeeding years by my colleagues Tracy Carrick and Darlene Evans. A librarian attends alternate class sessions, and students work with a special staff of undergraduate TAs to produce a twenty-page research project, which the students also present with illustrations during the final week of the semester. Our first finding was that even struggling writers can complete an ambitious project during their first year. A second finding was that the course was useful for students from a range of backgrounds with a range of abilities. A single class is not an array, however, and there are real limits on research writing in a regular FWS in the disciplines. Still, a lot can be learned in the freshman year. Each semester we join the librarians in judging entries for a prize for the best sequence of assignments in information literacy, to be awarded to a graduate instructor who teaches the FWS.

The Office of Academic Diversity Initiatives

In my first year as director of the Knight Institute, my chief goal was to make intensive writing courses more available to those students most at risk for failure in college. Locating some statistics deep in the university's database, I learned the percentage of target minorities who were failing to graduate and the gap between completion rates for target minorities and students who are "white." (In 2006 the largest figures were the dropout rates for African American males [25 percent] and for Native American males [50 percent]). The gap has narrowed since then, particularly for women, but students of color continue to earn lower GPAs than their peers at Cornell.

It made sense to assume that students from under-resourced high schools—students without computers or perhaps a private space to study at home—would be especially likely to trail their peers in writing skills. But obviously, the weakest writers are not all students of color, nor are all students of color from low-income families. For obvious reasons, the Knight Institute could not design courses exclusively for students based on income or ethnicity. On the other hand, we could offer courses in subjects that would attract students of color by means of the subject, and we would continue to offer the Writing Workshop seminars, designed for weaker writers. But the Workshop seminars are never mandatory; even our writing assessment exam, administered each fall, is by invitation. One result of this voluntary system is that the clientele for Workshop seminars now largely consists of international students. Students whose first language is not English are often eager to enroll in Workshop courses, whereas native speakers often wish to avoid the stigma of needing special writing instruction in their own language—especially in courses where most students hail from abroad and speak with accents.

We've never solved the problem of matching intensive writing courses with native speakers with weak writing skills or of serving students whose admission to college is a social as well as an academic challenge. At one point I proposed to join the Prefreshman Summer Program (designed to prepare students for their first full year in college) to a mandatory writing seminar similar to Writing and Researching in the University. But in order to build incentive into the program, students' counselors sometimes allowed the summer writing course to count as an FWS requirement. More recently, the university expanded its academic resources for underserved students, which suggests we'll have richer possibilities for collaboration. The new Office of Academic Diversity Initiatives has spacious quarters with room enough for many activities, including writing tutoring, and its director has secured several grants to bring gifted students from under-resourced urban areas to Cornell. It's possible to mandate courses for these students, and at the moment, a small group of scholarship students is slated to take Writing and Researching in the University—a welcome prospect but one far less ambitious than I'd once expected.

The Graduate School

Our staff has long been aware of graduates' need for tutoring—in some cases, fairly desperate need—because of the sheer number of requests we've received for help with writing theses and other research projects. But it's not easy for departments to understand why their graduate students, if competent, could need help. Surely a person admitted to a graduate program must have mastered the conventions and methods of his discipline. Only recently have educators begun to understand that graduate students struggle with writing in the same way that undergrads do when asked to write at a new level of sophistication—specifically, to write work of publishable quality in a discipline. The problem is not simply in the sciences; in fact, degree completion times among humanities grads are longer on average than those in the sciences—apparently because writing in the humanities is more solitary. Once again, Keith Hjortshoj broke pedagogical ground, this time by analyzing and defining the structures of graduate writing across disciplines. He consulted with individuals, taught a seminar called Works in Progress, and led a summer retreat (funded by a three-year grant from the Council of Graduate Schools) for graduate students with writer's block. Two of his books draw on these experiences (*Understanding Writing Blocks* and the Cornell in-house publication *Writing from A to B*).

Meanwhile, a proposal for a graduate tutoring service languished in a drawer in the Writing Workshop, waiting for funding and an interested party. Because of the intricacy of Cornell's administrative structure, the Graduate School hires no teachers and sponsors no courses, yet all graduate students are enrolled in the Graduate School, not in the college where their departments are housed. More recently, Tracy Carrick has trained graduate student tutors as part of her work for the Writing Walk-In Service. But tutors who work for the service are only expected to consult with students with modest needs—a single project, one or two conferences. In 2012 the dean of the Graduate School approached us to design a pilot tutoring program; Tracy quickly drew up a calendar for training a corps of tutors and a budget for paying them. If the funding continues, the tutoring service will remain a part of a group of initiatives.

The history of teaching international graduate students is longer and more complex. Its latest iteration began in 2013 when the provost gave a fixed amount to the Graduate School, renewable each year, to establish a program in graduate instruction in English as a second language (ESL) that would be administered by the Knight Institute, with the benefit of regular collaboration with its staff. The provost's allocation, though modest for a thriving program, was sufficient to hire a talented director, Michelle Cox, who in turn hired two staff members—the Knight Institute's newest colleagues. They form the core of the English Language Support Office (ELSO). For the first time in Cornell's history, the writing program and the Graduate School have assumed joint responsibility for providing language services to Cornell's international graduate students. In March 2015, Tracy's work with the tutoring service and ELSO's courses in English speaking and writing were recognized by an Award for Excellence and Innovation in Graduate Education from the Council of Graduate Schools. Among other interests, Michelle is now seeking ways to involve international students in the multicultural education of domestic students through collaborations between ESL instructors and WAC programs. She recently coedited a volume of essays that address this new form of informal collaboration (Zawacki and Cox).

University Courses

I mentioned at the start that Cornell's federal system of colleges holds out a promise of intellectual diversity that's partly hampered by mental habits and partly by rules that regulate teaching collaborations and the movement of students across colleges. To make learning across colleges easier, Laura Brown, the provost for undergraduate education, came up with the idea of a minicurriculum known as University Courses. Her goal was to adapt or invent a set of exemplary courses, which could be foregrounded in the Cornell catalog and other materials as courses that are particularly innovative and interdisciplinary. Some of these would be large, requiring TAs, and some would be writing intensive. For the original task force for University Courses, Laura chose several faculty members who had taught WIM courses, along with both of the directors of WIM, Keith Hjortshoj and Elliot Shapiro. Partly because of their familiarity with Keith and Elliot's work, faculty suggested that the Knight Institute be the administrative center of the new program. Elliot is now administrative coordinator of a program that mounts about twenty-four courses per year, many of them cutting across colleges and many of them team taught. He has also set up focus groups for each course and a series of self-study sessions that include the TAs. Thus, the new program bears the mark of two Knight Institute innovations: intensive training courses for graduate instructors and pedagogic conversations among faculty across disciplines. (As an example, a course on hip-hop included TAs from music, English, and Africana studies, and drew upon the world's largest collection of hip-hop materials, now housed in our Rare and Manuscript Collections. Writing assignments touched on such fields as urban sociology, ethnomusicology, and the visual arts.) In the latest fall orientation session, some four hundred new students attended presentations by faculty teaching these courses—a voluntary event that will probably replace the traditional (and less popular) freshman book project.

Writing with the Community

All education occurs to some extent through experience; all education to some extent involves knowledge about oneself. But service learning is preeminently about those things—experiences, the self—and writing is an indispensable means for converting experience into self-conscious knowledge that bears the potential of transforming a life. The theory of community-engaged writing frequently centers on the reflective journal or essay, which is understood to be the record or outcome of experiential learning. Through action and reflection, students often come to an awareness of their social position vis-à-vis a particular community and, by extension, of their position in the social whole—an awareness that in many cases has direct implications for students' own goals and values. This type of writing obviously departs, in some ways, from the norms of academic instruction and assessment. Reflective writing does not invent the university; it may strive to reinvent both the university and the self through new experiences of relationship.

The Knight Institute has no mandate and no funding for writing outreach. We've achieved what we have because of Darlene Evans, a dual career hire, who came to Cornell from a strong background of outreach work in Philadelphia schools. In her years as director of writing outreach, she's established and cultivated enduring partnerships with the Ithaca school system and with the Public Service Center. (My personal connection with the Cornell Prison Education Project is less formal, and I'll return to it at the end of this chapter.) In fall 2014 the Einhorn Family Charitable Trust announced a fifty million dollar gift to Cornell to expand its programs in local and global outreach, to be administered by the new Office of Engaged Learning. As these programs get under way, we will no doubt have fresh opportunities for pushing beyond the barriers separating academic learning and lived experience.

A first-year writing seminar centered in the public schools. In Darlene's seminar Finding Common Ground, Cornell undergraduates spend a semester with public school students as mentors or, in some cases, as collaborators in research projects on a topic of local interest. As with every well-designed service-learning course, Finding Common Ground seeks to break down the segmentation of our society—the invisible boundaries that, for example, block off the life course prescribed for children of privilege from other children who are discouraged, marginalized, or implicitly excluded. On the Cornell side, undergraduates typically develop a new understanding of the relationship between education and inequality.

The Public Service Scholars concentration. For decades, the Public Service Center has been Cornell's chief agency for placing student volunteers in locations throughout the community—as tutors in after-school programs, teachers of English for Spanish-speaking farmworkers, or visitors to a local detention facility for girls. Darlene teaches both the keystone course and the one-credit writing seminar for the new Public Service Scholars concentration, which selects from a group of gifted and highly motivated undergraduates and gives their outreach projects an academic context.

Polemic on Labor and Values in the Corporate University

Since I've been insisting that a good writing program depends in the first instance on a good workplace, I want to conclude by looking outward, from the privileged perspective of an Ivy League university, and putting Cornell's experience in the broad context of education in the twenty-first century. No one needs to be reminded that we've seen in our lifetimes the most profound and complex series of changes in the history of American universities, changes driven, for one, by the strategic defunding of higher education by federal and state government. Whole new, insecure populations have risen in the wake of this transformation: a generation of PhDs, particularly in the humanities and social sciences, who have little realistic prospects for a job; students burdened for decades after their graduation by massive loans with nondeductible interest; and casual educational laborers, now the majority of all faculty—as they are the overwhelming majority of writing faculty—working for low pay without the protection of tenure that once sanctioned free inquiry in the academy. The damage has progressed so relentlessly and ubiquitously that it's tempting, particularly for people in institutions protected (to some degree) by their affluence, to experience the process as a sort of background rumble, depressingly beyond human intervention. By the same token, the protracted crisis has produced a body of vociferous dissent that ruthlessly dissects the corporate university and advocates for change. In what follows, I begin with Marc Bousquet's *How the University Works* (2008) because it's a thorough-going attack on the corporate, or "managed," university that focuses on labor issues.

Mike-Crow Management

Bousquet analyzes the modern university as, quite literally, a corporation, composed of people at varying levels of management and labor whose class positions tend to determine their behavior. According to this schema, writing program administrators (WPAs) belong to lower management. In a corporate system, individual lower managers can sometimes identify with labor and sometimes with employers, but their general function is to represent the front lines of management, imposing and rationalizing managerial decisions from on high. According to Bousquet, "the disciplinary identity of tenurable faculty in rhetoric and composition has emerged in close relation to the permatemping of the labor force for first-year writing," creating what he calls a system of "tenured bosses and disposable teachers" (158). Yet these "tenured bosses" lack real power: they're "particularly vulnerable, highly individuated, and easily replaced," and their attempts to advocate for staff are generally ineffectual (159). His recommendation for the position of WPA is simple: it should be abolished. Instead of working as "disposable" labor managed by "tenured bosses," writing teachers should work in programs structured like academic departments, with a revolving department head rather than a WPA (163).[5]

Readers may easily find Bousquet's blunt language reductive or cynical; a good test would be how well his description predicts future institutional behavior. Last November, as I was reading his book, Arizona State University (ASU) raised the teaching load for writing faculty from four-and-four to five-and-five without a raise in salary. The payless raise was technically a conversion of the fifth part of a writing instructor's job (departmental service) to a fifth pair of courses, so university administrators were able to claim they hadn't increased workload. Non-tenure-track salary for teaching eight courses per year was $32,000, though a few months after attracting national attention for the payless job change, the administration raised salaries to $36,000 (a "whopping," as the popular press might put it, 13 percent) but kept the increased workload. It turns out that the enrollment cap for ASU writing courses is twenty-five, up from nineteen a few years ago; with a five-and-five teaching load, writing faculty will teach up to 125 students per semester—a load that's actually far from the highest in the country (Flaherty). It turns out that half the writing instructors are ASU PhDs, which at least indicates that these former grad students aren't unemployable, as so many others are; instead, their degrees become their ticket to low-wage, untenured jobs at their alma mater.

The decision to increase course loads came from the president, Michael Crow, but it's one of thousands of decisions this activist president has made since his arrival in Tempe (his "hands-on" administrative style has been nick-named Mike-Crow management). After the Arizona legislature slashed the university budget, Crow increased student enrollment from 55,000 to 82,000 while also raising tuition. Instead of adding faculty, he entered a lucrative partnership with Starbucks to launch an online education system; he also boasts of tripling the university's research spending. After the workload of writing instructors increased to five courses per semester, the trustees of ASU gave the president a 20 percent pay raise, to nearly $900,000. His next bonus ($90,000) will be linked to his success in achieving another targeted increase in undergraduate enrollment. Since no new faculty will be hired, existing faculty will teach the courses that aren't electronic, while the profits of the online collaboration go to Starbucks and the bonus goes to Crow.[6] As for the role of tenured faculty, either in decision-making or in protests against it, no trace is visible. (Although rumors of private anger abound on the web, I've found no public statements by faculty.) Meanwhile, Crow has branded his style of administration the "New American University," and a media campaign has been spreading the message, both of the man and his concept. *Time* named him one of the top ten administrators in the country ("America's Ten Best"); his coauthored book *Designing the New American University* was brought out by Johns Hopkins University Press and splattered with blurbs by high-profile political hacks (Bill Clinton, Jeb Bush) and by the senior education officer of the Gates Foundation. The *Chronicle of Higher Education* ignored the incident of the five-and-five course load, then devoted a celebratory cover story to his achievements in a spring edition of the *Chronicle Review*. The president's level of compensation produced a gush of justification by top-level management (the trustees) and the local media. "It's no secret that a lot of institutions would like to steal him away from us," one regent said, while the *Arizona Republic* pointed out that since the football coach earns twice the president's salary, Crow is actually underpaid. The newspaper didn't ask whether writing instructors at ASU are also underpaid.

What could a WPA say in such a situation, and how could she advocate for her staff—in this case, the largest writing staff of any American university? Very little, it would seem, given the power of management at ASU, even though the new workload is in clear violation of the "Principles" of the CCCC—of which the WPA at ASU is a past officer. In the fall 2014 online newsletter of ASU's writing programs, *Writing Notes*, the director focused her response not on the question of workload but on what she correctly calls the "closed system" of ASU PhDs—and then offers this explanation: "Across the country, PhDs and MFAs in English are taking non tenure-track full-time teaching jobs because this is the way the work of higher education teaching is being restructured."[7]

In this case, we are all being restructured—in the same sense that an injustice to one is an injustice to all.

If Bousquet's analysis appears to predict quite well the state of affairs at ASU, how would it apply to Cornell? In the case of graduate student labor, since TAs at Cornell teach one course per semester with a full package of support, I believe their labor constitutes valid professional development and not in our case exploitation. (In principle that is true. I'm putting aside the fact that in the present economy, many PhD holders will not get jobs, and that becomes a discussion about reducing or eliminating graduate programs.) As far as the director's position is concerned, it would be ironic if an essay that began by praising the structure of the Knight Institute's program wound up consuming itself, with concluding that my own job—the position that's given me the vantage point to write—is superfluous in the best of all possible worlds (or even in the present imperfect world of Cornell). Nevertheless, I find Bousquet's argument for abolishing the WPA—or more to the point, for elevating a writing program to the status of an academic department—to be correct. Several of my colleagues are already administrators (as well as published researchers), and it's obvious that staff members could easily rotate as chair, as Bousquet recommends—taking on the responsibility for planning budgets, advocating for students and resources, and representing the program to other faculty and to administrators. Only the requirement of symbolic capital—the prestige that is the common coin of the academic economy—justified the decision at Cornell to make the WPA a professor. At the same time, the structure of the Knight Institute probably allows the relationship between WPA and staff to be as close to equality as is possible outside the structure of an academic department. The sticking point is tenure. As an Ivy League institution, Cornell does not hire PhDs in rhetoric and composition at the professorial level, which means that we stand to lose qualified candidates in future job searches. I've pointed out that we do have "tenure without tenure," which has historically amounted to job security in almost all instances. (In the budget crisis of 2010, the English department in fact had to let two senior lecturers go, since their unguaranteed salaries constituted the only place where the department could shrink its budget.) Yet the entire thrust of this chapter has been to argue that the Knight Institute's writing "nonspecialists" have an expertise that breaks down the outmoded distinction between research positions (tenured) and teaching-only positions (untenured). In their recent book, Michael Bérubé and Jennifer Ruth argue forcefully for granting tenure to full-time teachers in an English department under certain circumstances; their argument holds equally well for my colleagues in a WAC program.[8] As I've also mentioned, Cornell does not do casual hiring—unlike such other well-endowed institutions as Harvard and Yale—except in one limited category (instructors who can be hired for one year only without a search). Assuming my job title will remain for the foreseeable future, I offer the following manifesto, keeping the hat I presently wear and not exchanging it for another. My second conclusion moves from the workplace back to the content of teaching.

First Conclusion: Manifesto of a Tenured WPA

The corporatization of higher education in America—for students, faculty, and the pursuit of knowledge itself—is a social catastrophe; the collusion of faculty in that process is a moral catastrophe. If the primary purpose of tenure is to protect freedom of speech and thought—and not simply to produce a privileged social class—then tenured faculty, guided by honest expertise (including knowledge of the structure of their own workplace), are *licensed to seek truth and also to deploy it*—to make it public and active. I've lately wondered what an equivalent of the Hippocratic oath might look like if tenured faculty, including WPAs, had to take one. This might be a task for the Executive Committee of the CCCC. The alternative to speaking truth to power is, by not speaking, to embody the untruth of power. I'm not sure, as some have suggested, that faculty can lead a coalition of activists, but I do expect to support efforts by casual academic workers to organize; to sign petitions and write letters and make donations; to join the noisy public actions that are bound to increase—an easy thing to do, admittedly, for someone in my position. Professional organizations should use the continuing crisis to frame conferences; to take public stands and raise funds; to be ready to expose egregious examples of institutional misconduct to public contempt; to support WPAs who are determined to resist the further exploitation of their staff; and perhaps to generate a more binding equivalent to the Hippocratic oath than this one. If for no other reason, tenured faculty need to insist on fairness in the academic workplace to be ethically consistent when defending liberal education. The two issues are of course connected, both positively and negatively: high standards of employment correlate directly with high standards of teaching and learning, whereas the blasts of "market fundamentalism" are an ethical assault, both on workplace justice and on the value of knowledge learned and enjoyed for its own sake.

Second Conclusion: The Right to Social Knowledge

Recent defenses of liberal education in terms of learning outcomes contain a paradox. By stressing intellectual skills rather than quantifiable knowledge—virtues like integrative learning, critical thinking, intellectual adaptability, integrity, and the like—those who would hope to rescue higher education from the purely instrumental thinking of the corporate university find themselves speaking the same language as corporate recruiters. The point is not that the virtues of flexibility, creativity, and the like are wrong, only that they can be reduced to individual traits that can then be marketed for any purpose—which is far from the kind of social good envisioned by the Morrill Act. The list of skills and outcomes leaves out what I here call the right to social knowledge: our students' chance to see the divisions and exclusions of the social world as part of a profound unity with moral imperatives implied. The idea of interconnections brings me back to the fundamental premise of a WAC program, staffed by people who are adept at forming conversations at the interstices of traditional boundaries between skills and forms of knowledge.

A friend of mine who is an aerospace engineer told me that when he came to his new job at Cornell's College of Engineering, he asked who taught the course on war. That course was never taught, at least not by the engineers, and Cornell may never get there. But it was a good thought. It's obvious that human survival will depend on the human ability to integrate scientific knowledge, environmental justice, and social need, and that the present and the future will need both STEM and flower: scientists who are political activists and poets who understand experimental evidence, historians of technology and engineers with a sense of the past, doctors with a moral commitment and philosophers fascinated by practical problems. At a superficial level, everyone today is in favor of interdisciplinarity; what I'm pointing to is not simply a learning outcome but a glimpse at the social realities in the real world that are somewhat reflected in the interdisciplinarity of the academic world; I'm pointing to understanding how discourses are used in the social world to clarify those connections or to distort them. A public sphere in which politicians can get elected by expressing skepticism about evolution and global warming is not a space where science is respected but one where the criteria of belief are systematically confused and undermined. Citizens of the world will need to grasp the dimensions of environmental change and learn to exert political will; policy planners will need to speak to climate scientists, and both to medical professionals, and all three to experts in urban affairs and others, and everyone to speak across nations and languages. We'll need, if not a single common language, then at least a degree of mutual intelligibility and a sense of common interests. "Liberal education" may or may not be an adequate phrase for the *democratic* values (to use a word that's curiously absent from the liberal discourse of learning outcomes), and for the willingness to struggle together, that our students will need if they are to improve the world. A university's experts on words and meanings will surely have a place in preparing them for an urgent future.

The idea of interconnections also brings me back to a collaboration I mentioned briefly above, the Cornell Prison Education Program (CPEP). Since the course I'll now describe is one I teach, this essay will end on a personal note.

The Morrill Act specifies that a land-grant college must justify taxpayers' investments by giving back to its surrounding communities the fruits of its research and expertise. The inmates at Auburn Correctional Facility are for legislative purposes residents of the county contiguous to Ithaca's. The CPEP offers courses taught by Cornell faculty and graduate students to incarcerated men, who are then able to earn associate's degrees. For three years, this growing program—the largest of the service-learning opportunities available to Cornell students—shared the office suite of the Knight Institute. In 2011, as a teacher in the program, I developed the service-learning course Writing Behind Bars for undergraduate volunteers. They don't receive credit for their work in the prison, but they can receive up to four credits for the course, which assigns weekly reflection essays on their experiences in prison and readings on such topics as literacy, inequality, and the politics of policing and incarceration. Seniors in Writing Behind Bars often report that their visits to the facility are the most valuable part of their undergraduate education; several have also said that their final papers "connected the dots" of their last four years. For Cornellians, the drive north from a sheltered campus in Ithaca and past the gates and through the courtyard of the nation's oldest prison breaks through the barriers—symbolically and almost literally—that in American society wall off rich from poor, professionals from laborers, black and tan from white, rural from urban, red from blue, those who afford college and those who don't, those who rise to the boardrooms and those who sink to the isolation cells. Though there are no formal ties between the Knight Institute and CPEP—between McGraw Hall on Ezra Cornell's hilltop and the barred palace in downtown Auburn, New York—CPEP links the two institutions in a mirror relationship that adds a new dimension to the teaching of writing at Cornell.

Notes

1. The Morrill Land Grant Act of 1862 (Thirty-Seventh Congress, session 2) specified that the colleges built from the sale of public lands will teach (among other subjects) "such branches of learning as are related to agriculture and the mechanic arts . . . in order to promote the liberal and practical education of the industrial classes in the several pursuits and professions of life" (chapter 130). The standard history of Cornell in the earlier years is Morris Bishop's *A History of Cornell*.

2. The most accurate history of the early years of the program, beginning in the 1960s, is Katherine Gottschalk's "Putting—and Keeping—the Cornell Writing Program in Its Place: Writing in the Disciplines." Essays by faculty members who have taught in the Knight Institute program have been collected by Jonathan Monroe in *Local Knowledges, Local Practices*.

3. Writing 7100 meets once a week for six weeks, either in the summer or fall, in sections staffed by Knight Institute lecturers and cofacilitated by experienced graduate instructors. Graduate students who take the course in the summer develop their fall syllabi during the last sessions of the course; they also train as classroom interns for Cornell's summer school, which is largely open to high school students seeking precollege credit, so that by the time the fall begins, the new instructor has helped conduct classes and completed a syllabus. The course works in the first instance to ensure uniform teaching standards across departments. Our writing pedagogy emphasizes frequent low-stakes writing, frequent revisions, logical assignment sequences, and the integration of writing into course content. We recommend that about one-half of class time be devoted to writing. Among other texts, we use two by colleagues at Cornell that have sustained national reputations: Gottschalk and Hjortshoj's *Elements of Teaching Writing* and Hjortshoj's *The Transition to College Writing*.

4. The sentences in the CCCC's 2013 "Principles" seem to have been constructed so as to excise any reference to reading with a scalpel, in some cases by substituting awkward phraseology for the *R* word. For example, the eighth principle, which concerns WAC, refers to the content of disciplines as "information, ideas, and argument"—suggesting that students can be exposed to legible material in a writing classroom but that writing somehow always takes place on its own. The one instance of the word *reading* I could discover occurs in the ninth principle, which refers to the preparation students have had *before* attending college: "Students come . . . with a wide range of writing, reading, and critical analysis experiences." The phrase describing the content of technological tools is "forms of composed knowledge," which can include podcasts

and videos. (The CCCC's position statements are available on its website: http://cccc
.ncte.org.) Once again, Hjortshoj's essay is pertinent: "If we hope to remain useful as
writing specialists[,] . . . we need to maintain agile, interdisciplinary, independent
connections with the essential living medium of all academic life, language itself"
("Alternative History" 83).

5. Since Bousquet wrote these words, adjuncts have continued to organize as members of the Service Employees International Union and other unions and have achieved substantial gains in salary and job security—partly because administrators are realizing that annual hiring has its own costs. (In fact, adjunct faculty are slightly declining in numbers; the real "growth category" is full-time non-tenure-track instructors.)

6. The blurb by the Gates Foundation official is as follows: "This book is a route map, rationale, and guide for the few and the bold who dare to step forward to build the universities we need for the twenty-first century." In the context of non-tenure-track salaries, the copy prepared for the dust jacket by Johns Hopkins University Press is Orwellian in the strict sense defined by the author of *1984*: "Crow has led the transformation of ASU into an egalitarian institution committed to academic excellence, inclusiveness to a broad demographic, and maximum societal impact" (qtd. in Warner).

7. ASU's English department runs many graduate degree programs. According to a blog writer who studied the English department's placement figures,

> In light of these numbers, I'm looking for a good reason for the Arizona
> State PhD program in English to exist, other than to create a continuous
> supply of low-paid non-tenure-track faculty and graduate assistants to teach
> the general education undergraduate courses in English at Arizona State,
> 50% of which are staffed by their own alumni. (Warner)

The "Principles" of the CCCC specify "teaching loads and class sizes that are consistent with disciplinary norms" as well as "paying instructors a reasonable wage and providing access to benefits." The document used to specify a maximum teaching load of sixty per semester; according to our conference colleague Doug Hesse, sixty should still be considered a good guideline. The Association of Departments of English recommends no more than three composition courses per semester.

8. Ruth and Bérubé predicate their argument in favor of tenure on the striking notion of a right to democratic participation in university governance, which they maintain is a useless right without the protection of tenure. They include among tenurable employees only those hired from a formal search. In 2014 Cornell's College of Arts and Sciences created the rank of professor of practice for certain nonresearch positions, but this applies only to people without PhDs who have achieved distinction in a particular field (a creative writer, for example).

Works Cited

"America's Ten Best College Presidents." *Time*, content.time.com/time/specials
/packages/article/0,28804,1937938_1937933_1937917,00.html.
Arnold, Matthew. *Culture and Anarchy*. 1865. Edited by J. D. Wilson, Cambridge
UP, 1971.

Bartholomae, David, "Inventing the University." *Journal of Basic Writing*, vol. 5, no. 1, 1986, pp. 4–23.

Bérubé, Michael, and Jennifer Ruth. *The Humanities, Higher Education, and Academic Freedom: Three Necessary Arguments*. Palgrave Macmillan, 2015.

Bishop, Morris. *A History of Cornell*. Cornell UP, 1862.

Bousquet, Marc. *How the University Works: Higher Education and the Low-Wage Nation*. New York UP, 2008.

Carillo, Ellen. *Securing a Place for Reading in Composition: the Importance of Teaching for Transfer*. Utah State UP, 2014.

Flaherty, Colleen. "One Course without Pay." *Inside Higher Ed*, 16 Dec. 2014, www.insidehighered.com/news/2014/12/16/arizona-state-tells-non-tenure-track-writing-instructors-teach-extra-course-each.

Gottschalk, Katherine. "Putting—and Keeping—the Cornell Writing Program in Its Place: Writing in the Disciplines." *Language and Learning in the Disciplines*, vol. 2, no. 1, 1997, pp. 22–45.

Gottschalk, Katherine, and Keith Hjortshoj. *Elements of Teaching Writing*. Bedford/St. Martins, 2003.

Hjortshoj, Keith. "An Alternative History of an Interdependent Writing Program." *A Minefield of Dreams: Triumphs and Travails of Independent Writing Programs*, edited by Justin Everett and Cristina Hanganu-Bresch, UP of Colorado, 2017, pp. 63–84.

———. *The Transition to College Writing*. Bedford/St. Martin's, 2001.

———. *Understanding Writing Blocks*. Oxford UP, 2001.

———. *Writing from A to B*. Cornell University, 2010.

Jamieson, Sandra. "What Students' Use of Sources Reveals about Advanced Writing Skills." *Across the Disciplines*, vol. 10, no. 4, Dec. 2013, wac.colostate.edu/docs/atd/reading/jamieson.pdf.

Monroe, Jonathan, editor. *Local Knowledges, Local Practices*. U of Pittsburgh P, 2016.

Rose, Shirley. Director's notes. *Writing Notes*, vol. 15, no. 1, Fall 2014, Writing Programs, Department of English, Arizona State University, english.clas.asu.edu/sites/default/files/writing_notes_fall_2014.pdf.

Warner, John. "ASU Is the 'New American University'—It's Terrifying." *Inside Higher Ed* blogs, 25 Jan. 2015, www.insidehighered.com/blogs/just-visiting/asu-new-american-university-its-terrifying.

Zawacki, Terry Myers, and Michelle Cox, editors. *WAC and Second Language Writers: Research towards Culturally and Linguistically Inclusive Programs and Practices*, Parlor Press / WAC Clearinghouse, 2014.

6

At the Intersection of Feminism, Rhetoric, and Writing Program Administration

Cheryl Glenn

Characterizing writing programs as recognizable organisms, generated, shaped, and individualized within a specific programmatic context and by specific personalities, Glenn speaks to the powers of feminism and rhetoric at the site of writing program administration before homing in on her own location at the Pennsylvania State University. Her own "notes toward a politics of location" (Adrienne Rich's phrasing) contrast with those of other feminist administrators. Like Jeanne Fahnestock's, Teresa Redd's, and Kelly Ritter's contributions to this volume, Glenn's chapter traces the history of a university-wide writing program. Along this historical trajectory, she inserts scenes of feminist and rhetorical contributions to administrative, curricular, and pedagogical decisions. Finally, she reflects on the powers and limits of writing program administration inflected with rhetoric and feminism.

"With whom do you believe your lot is cast? / from where does your strength come?" (Rich, "IV" 6). So asks Adrienne Rich, one of the many feminists with whom I have cast my lot and from whom I borrow strength as a human, a teacher, and an administrator. Rich first appeared on my radar when I discovered she had taught alongside Mina P. Shaughnessy. Led by Shaughnessy, one of the most influential writing program administrators (WPAs) our field has known, Rich and many other activist women launched a basic writing program at the City College of New York.[1] As a response to open admissions, these women collaborated with one another, reached out and listened to smart students who had been ignored, and changed the face of higher education at a time (late 1960s, early 1970s) when the civil rights movement in America truly meant positive change.[2] Like the best of rhetors, the best of feminists, the best of administrators, these women made their rhetoric—their words, silences, pedagogy, and actions—*do* something.

As a collective force, these women represent the many ways to move the feminist rhetorical project into the twenty-first century, carrying forward its initial goals of recovering female historical figures and female-authored literary works (Glenn, *Rhetoric Retold*; Logan; Lunsford; Royster) at the same time that it seeks out social justice for both women and men, just as Shaughnessy had done with her basic writing program. Those possibilities for transforming the status quo into social change are realized by feminist rhetors, writers, teachers, and administrators, many of whom deliberately implement nontraditional rhetorical practices. In individualistic ways, they strive to engage in authentic dialogue, straightforward deliberation, and productive collaboration, all the while tapping such feminist values as invitation (Foss and Foss), listening (Lipari; Ratcliffe, *Rhetorical*), and productive silence (Glenn, *Unspoken*).

In this essay, then, I speak specifically to the powers of feminism and rhetoric at the site of writing program administration. I open by plotting women's settlements on the landscape of writing program administration before homing in on my own location at the Pennsylvania State University. Inevitably, my own "notes toward a politics of location" (to use Rich's phrasing) will contrast with those of other feminist administrators. Like Jeanne Fahnestock's, Teresa Redd's, and Kelly Ritter's contributions to this volume, mine traces the history of a university-wide writing program. I map the rhetorical history of Penn State's writing program, illustrate its current iteration, and chart its probable future. All along this historical trajectory, I insert scenes of specific feminist and rhetorical contributions to administrative, curricular, and pedagogical decisions. Finally, I reflect on the powers and limits—the rewards and frustrations—of writing program administration inflected with rhetoric and feminism.

Across the Nation

Listen to the women's voices. Listen to the silences, the unasked questions, the blanks. Listen to the small, soft voices, often courageously trying to speak up.
—Adrienne Rich, "Taking Women Students Seriously"

For most of the history of writing programs, women have held relatively few administrative positions, serving, instead, as instructors with limited input into the programs' philosophy and curricula. So in writing this essay, I join the comparatively recent cohort of feminist WPAs who relate their experiences within the context of their location. Many of these eloquent and truth-telling administrators are beleaguered, if not powerless, overseeing a cadre of equally overworked, often underappreciated writing instructors, most of them women.

Feminist administrators and scholars alike have described much the same scene. Early on, in her 1991 "Women's Work," Sue Ellen Holbrook argued that the professionalization of composition (1870–1930) was actually a "feminization" of composition:

> [I]n the status hierarchy within English, teaching was inferior to research . . . ; work with undergraduates, especially freshmen, was inferior to work with graduate students; and composition . . . was inferior to the scholarly, scientific conception of literature. Hence, in the new university, composition teaching began positioned in the lower status, and there it has remained . . . as appropriate work for paraprofessionals—and for women. (207)

In *Textual Carnivals*, longtime WPA, scholar, and teacher Susan Miller interrogates the status of composition in the university, describing the field as the "Sad Women in the Basement." In her chapter by that same name, she taps the assessments of compositionists themselves to rank their field in relation to literary studies, critical theory, and rhetoric; to position their academic ranks (fixed-term or tenure-accruing) and expectations of promotion; and to locate compositionists in the larger public sphere, what with its surplus of PhDs in English studies writ broadly. Pulling from Peter Stallybrass and Allon White, S. Miller describes the compositionist as "an underground self with the upper hand" (Stallybrass and White 4), someone with a contradictory identity, living in the academic basement yet rousing public authority:

> It is an employment that in the majority of its individual cases is both demeaned by its continuing ad hoc relation to status, security, and financial rewards, yet given overwhelming authority by students, institutions, and the public who expect even the most inexperienced composition teacher to criticize and "correct" them in settings entirely removed from the academy. The perduring image of the composition teacher is of a figure at once powerless and sharply authoritarian. . . . (139)

As Eileen Schell explains in *Gypsy Academics and Mother-Teachers*, composition's embrace of feminism, what with its values of "nurturance, supportiveness, interdependence, and nondominance," works mostly to normalize writing instruction as "women's work"—neither serious, rigorous, nor intellectual (76). Of course, being coded female or feminine in the workplace is rarely good; it means being relegated to the basement, the subaltern, the margins.[3]

Theresa Enos, another longtime WPA, expands S. Miller's work, explaining that the administrative work in writing programs stretches beyond directing the program itself. Such administrative work includes serving as a director or coordinator[4] of a specific writing course (first-year writing, business writing, and so on); of the writing center, lab, or clinic;[5] or of various programs, including English as a second language, peer tutoring, teaching assistant training, computers and composition, English education, summer session, assessment, and so on. Notwithstanding the fact that many rhetoric and writing specialists are involved in the full range of departmental administration (from supervising peer tutoring to serving as department chair), the women in this administrative cohort report that they are "paid less and work harder" than their male counterparts and their counterparts in literature (Enos 72).

Given that the term *research* is considered to be "masculine," as are the terms *rhetoric, theory*, and *graduate seminar*, and that the term *teaching* is considered to be "feminine," along with such terms as *composition, undergraduate curriculum*, and *service*, it should come as no surprise that Enos uses the phrase "female ghettos" when describing most writing programs, where from 60 to 90 percent of the instruction is performed by women, whether in tenure-track, adjunct, or fixed-term positions (78). Thus, the gendered politics of writing instruction and WPA work continues to mar the academic scene.[6]

WPA Sally Barr-Ebest reveals four reasons why female WPAs experience disempowerment in terms of respect, salary, and promotion, all of which she lays at the feet of deans and department chairs who cannot distinguish between service and administration, remain uninformed about the complexities of writing program administration, do not accommodate WPA responsibilities within tenure-and-promotion guidelines, and are influenced by unconscious sexism and socialization (63). In fact, her respondents (male and female) provide her a list of gendered problems in WPA work that can be summarized as "women take responsibility, men take authority" (66). Some of the problems include "women work harder for fewer rewards, status, and their personal lives are more difficult," and "women are used for the hard, time-consuming work but not rewarded with power" (66).

Or, as Lynn Z. Bloom so wryly sums up the gendered problems associated with WPA work,

> I want a Writing Director who will want to remain a Writing Director for the rest of her days, and who will find fulfillment in this most ennobling, if humbling, of tasks. Once I have shown her the ropes I will expect her to handle everything; we will indeed be a team, but I will be the titular head, the silent partner. . . . (177)[7]

Bloom continues, in the vein of Judy Syfers's celebrated feminist declaration "I Want a Wife," and concludes, "My God, who *wouldn't* want a Writing Director?" (178).

Drawing on the most influential WPA-related research from the past twenty years, Jonikka Charlton and Shirley K. Rose extend the exploration into the value of a WPA by charting "Twenty More Years in the WPA's Progress."[8] The authors observe some measure of improvement in the status of WPAs, in general, and of female WPAs, in particular, as well as in the future of WPA work itself (138). Nevertheless, Charlton and Rose reject the idea that the Council of Writing Program Administrators has "somehow arrived at our Celestial City" (115). Likewise, coeditors Krista Ratcliffe and Rebecca Rickly remind us that, as we settle into the twenty-first century, feminism has made fewer inroads into WPA work than earlier writers such as Holbrook and S. Miller might have hoped.

In nearly thirty years, the material conditions of the work simply have not changed much. Whether male or female, feminist or not, all WPAs continue to work within an academic-administrative-organizational hierarchy with mandates for reporting (and being reported to), delivering and receiving information, and getting things done, making our rhetoric do something. After all, Rich reminds us, "the university is above all a hierarchy," one "built on exploitation" ("Toward" 136, 145). The university-wide requirement that is composition affects everything—from the bursar's and admissions offices to academic advising, writing program staffing, and adjunct hiring. Even though the contributors to the Ratcliffe and Rickly collection invoke feminist practices of caring or nurturing, collaboration, active mentoring, being interrupted, attention to gender issues and process, and distributed responsibility, these practices are all submerged in the quotidian pressures of getting things done in what the editors refer to as the "administrative trenches" (213). Little wonder, then, that contributor Nan Johnson would raise the issue of the "troubled intersections of feminist principles and administrative practices" (qtd. in Ratcliffe and Rickly 213).

Johnson's point is the crucial one: What difference do feminist principles actually make within such an academic-administrative arrangement? Many of the essential, ideal qualities of a feminist WPA are impossible to employ when taking an administrative position, whether the WPA is experienced or not, tenured or not. Each administrative position is contained by or contains its own scene. No one can enter the scene fully prepared; there is always on-the-job training as one learns the ins and outs, the backstories and the histories of any scene. As Kathleen Blake Yancey puts it so succinctly, "We all work in a context" (153). We each invoke and embody our feminist principles within our own location.[9]

Location, Location, Location

*I have been working to change the way I speak and write, to incorporate in the
manner of telling a sense of place, of not just who I am in the present but where I
am coming from. . . . I refer to that personal struggle to name that location from
which I come to voice—that space of theorizing.*

—bell hooks, *Yearning*

It has become de rigueur in feminist circles to announce one's standpoint, one's
location, before embarking on a narrative, a theoretical analysis, or epistemological exploration. From Nancy Hartsock and Sandra Harding to Patricia Hill
Collins and Adrienne Rich, scholars write about the significance of location
to one's epistemological consciousness as well as to one's understanding (or
lack thereof) of power relations. All of these women invoke sex, phenotype,
socioeconomic status, level of education, and religion (among the many markers of identity) as constituting material locations of us all.

I am called on to do the same as I describe my status vis-à-vis the well-
established, rhetorically based writing program of Penn State. A white, middle-
aged, heterosexual, female, tenured professor of English, rhetorically trained
at one Big Ten university and fully employed at another, I spent fifteen years
at Penn State (preceded by eight years at Oregon State University) before I
agreed to administer the huge writing program. Unlike many WPAs, such
work was neither part of my graduate training nor my professional identity.
And unlike so many other WPAs, I had a choice.

By 2012 I felt comfortable identifying myself as a professionally established
scholar, writer, teacher, and leader. And, of course, I was older. By that time,
I knew I could tap my broad administrative and leadership experience, having
held presidencies and served on the executive committees of numerous professional organizations, directed several programs, established two university
centers, taught a wide range of courses, and worked through the inevitable progression of thorny (occasionally heartbreaking) issues that accompany anyone's
personal and professional life. I was established when I agreed to take on a big
administrative commitment. I was not saddled with any expectations of serving
as caretaker for a predecessor's agenda (as is the situation in so many places);
rather, I could build upon, extend, or otherwise rethink the rich contributions
of my mostly male predecessors. At least that is how I perceived my location
then—and how I perceive it now. Thus, my account of writing program administration bears the ideological weight of my own body, standpoint, location,
mesosystem. I cannot reduce that weight, but I can acknowledge it.

Writing Program Administration at Penn State

The questions that we have to ask and to answer about that [academic] procession during this moment of transition are . . . do we wish to join that procession, or don't we? On what terms shall we join that procession? Above all, where is it leading us, the procession of educated men?   —Virginia Woolf, *Three Guineas*

Directing what is now called the Program in Writing and Rhetoric (PWR) is arguably the biggest administrative responsibility in the English department after that of the department head.[10] The number of students and instructors alone complicates the WPA work. At University Park,[11] fifteen thousand undergraduate students move through our approximately seven hundred sections of writing and rhetoric courses each year, taught by over sixty graduate students and approximately one hundred fixed-term lecturers (or adjuncts).[12] Additionally, the position of WPA carries with it the responsibility for two university-wide requirements, necessitating administrative contact across, up, and down organizational ladders across all twenty-four campuses.

Built on a strong foundation laid by Fred Lewis Pattee, father of American literature scholarship, and then fortified much later by the rhetorical and pedagogical expertise of University of Chicago's Wilma Robb Ebbitt, Penn State's PWR continues to flourish. The PWR comprises a wide range of courses, from developmental writing[13] and various formulations of first-year writing that include a one-semester thematic honors course to a two-semester honors track in writing *and* speaking.[14] PWR courses also include upper-division writing courses in various disciplines and upper-division rhetoric courses—all rooted in rhetorical principles.[15] The pedagogical goal is to help all our students—not just the smartest, richest, or best prepared—become strong, flexible, confident writers, capacious in their understanding and application of rhetoric. They learn to write alone, draft after draft, as well as in concert with their instructors and their peers, delivering their compositions in words, images, and multimedia. In addition to being responsible for all the instructors in the PWR program, I am also responsible for the credit-based courses and tutorials that constitute the undergraduate writing and graduate writing centers.

Putting Feminism to Work

A feminist vision of personal power is likely to be quite different. It represents a different way of exercising power because it is based on a different notion of what power is. At base, power is seen as a limitless rather than finite quality.

—Hildy Miller, "Postmasculinist Directions in
Writing Program Administration"

My recent work with Penn State's Writing in Business course, our most populated upper-division, discipline-specific writing course, reveals the administrative issues (hiring, budgeting, online instruction) faced by nearly every WPA. At the request of our Smeal College of Business, I was asked to create reserved sections of our business-writing course for Smeal juniors, who were frustrated because they could not enroll in this required course during the semester they preferred.[16] The Smeal administrators asked me to "do something." As a feminist rhetor, I tried to make my words *do* something akin to Kate Ronald's "bold vision" (160). And like Kate's, my feminist rhetorical vision, too, was a negotiated one.

I needed to add sections right away, and I leveraged this exigence as means for making additional hires of specialists in business writing. Although I was initially advised by my associate head and associate dean that we should simply use our (very good but already overcommitted) fixed-term lecturers to meet this need, I was able to convince them that we should try to hire PhDs at the rank of visiting assistant professor (VAP) by framing the discussion with "courses worthy of Penn State students" (more about that phrase below).[17] In the end, we hired three VAPs of business writing, new PhDs with specialties in business and professional writing.[18] At competitive assistant professor salaries, these three VAPs taught a three-three load of sections reserved for Smeal students during their two-year contracts before moving elsewhere. Despite the fact that various administrators had run calculations to figure out exactly how many sections of business writing we needed to offer to meet the demand, we did not meet the demand. We are hiring again—"up to six" lecturers or senior lecturers for business writing with three-year contracts.[19] I am not sanguine that this current effort will completely resolve the course-registration problem, but, then, I do not have control over what times and days Smeal students will take their classes or who will pay for their business-writing instruction. Still, I am pleased that this feminist-inflected process leads to the hiring of these productive colleagues.

My preference is to hire a tenure-track professor (or two) in business writing who can help us prepare even more graduate students and lecturers to teach this course, along with technical writing and courses in professional communication, both residential and online. But, in my location, the director of the PWR does not control the budget and cannot, therefore, move line items around to accommodate tenure-track hires, even those that I feel would enhance the program in a number of ways. Hiring, budgetary, and online instructional demands all constitute major issues for the future of the PWR.

In "Defining Moment," Yancey speaks to the importance of "taking hold of the budget" for WPAs and, especially, for female WPAs (149). I think she is right for many reasons: understanding how the money "works," how budget lines can be realigned with priorities, and how combining responsible planning with agency can all produce positive results for individual units. Yet the material conditions of the PWR do not easily lend themselves to such a model: its directorship has always been a rotating position; the PWR's hiring needs have always been determined by the registrar's office;[20] and our dean supplies the funds for our fixed-term appointments. Our dean also controls the tenure lines, after our department has voted on hiring priorities.[21] And given that we remain a comprehensive English department, the PWR's hiring needs and budget remain inextricably intertwined with those of literary studies, creative writing, critical-cultural theory, visual studies, and other emphases. It is within such a context that I work to conceptualize how best to meet the demands of all our courses but especially our upper-division courses and our online curriculum. As Linda Alcoff reminds us, "We must . . . interrogate the *bearing of our location and context* on what we are saying" (112), and I do, as I think through the daily and future demands of hiring, courses, and curriculum.

How to face those demands as a WPA employing feminist rhetorical practices is an approach I take seriously. So with Yancey's good counsel always in the back of my mind, I apply the feminist models of WPA work put forward by Lynn Z. Bloom, Marcia Dickson, Amy Goodburn and Carrie Leverenz, Jeanne Gunner, Laura Micciche, Hildy Miller, and Eileen Schell, all of whom critique what has been called the WPA-centric work model of centralized power in favor of one that entails a WPA's willingness to "relinquish control over the world" (Gunner 152). These feminist WPAs resist models of administrative work that centralize power in one individual (Micciche 449). In response to that traditional model (which would, for good reasons, include control of the budget and hiring), both Gunner and Micciche invoke the importance of rotating the position, distributing power and responsibility, establishing productive committees, and rewarding excellence and effort. I would add to their feminist rhetorical model the ability and willingness to reach out, collaborate, negotiate, strive for mutual understanding (rather than persuasion), and work strategically toward a vision.

Wisely, H. Miller cautions that such "leadership can appear weak if receptivity is mistaken for passivity; affective responses such as laughter for lack of seriousness; and the sharing of power for looking to others for direction" (54). And she goes on to remind feminist rhetorical WPAs that "one person at the top must function as a figure to take both credit and blame. As a result, . . . feminist directors must alternate feminist and masculinist personas to cope with double ideologies . . ." (56). In other words, I must be able to lead purposefully and responsibly and also be able to collaborate and negotiate toward a good (or good enough) result. Coupled with H. Miller's advice, the feminist rhetorical model for a WPA put forward by this group of women aligns better with the material conditions of my position than Yancey's good model does; it aligns better with my mesosystem, including the cultural and administrative structure within which I work (from Penn State "Pride" and East Coast students to the decanal and headship structures and my English department colleagues). Of course, this feminist rhetorical model also seems to align with my own disposition as well, with the ways I have found to interact most productively with my colleagues, again and again.

Facing the Future

It's our future, so we'd better be there.   —Susan Gubar, *Rooms of Our Own*

Facing the future as director of the PWR includes establishing inclusive, power-distributed ways to guide changes in our courses, hiring, and teaching assignments within a national culture of ever-reduced budgets, ever-expanded leveraging of media, and an increasingly wide gap between underpaid, underemployed adjunct labor and faculty. All of these issues come into play as I move into the future of the Penn State curriculum. Fortunately, feminist rhetorical principles and practices continue to sustain me in my work.

A few years ago, the faculty senate launched a general education review that threw the entire campus in a panic, as the senate wanted to make change— and fast. The last review of Penn State's general education requirements had been some twenty years ago, in 1995, so it seemed appropriate to review them again. Doing so could serve as a welcome opportunity to revisit which courses are required and what required courses should achieve. But because no one in writing studies had been included in the initial discussions, I initially resisted responding to requests for approval so that "they" could not make changes fast. Fortunately, no one else in English was interested in the rumored changes either, changes that included a "thematized and clustered" undergraduate curriculum that would begin with a seven-themes approach to first-year writing—or so rumor had it.[22]

I found that the best means for inviting productive, mutually empowered conversations about the curricular review (and thereby slowing down the proposed changes to first-year writing) was to ask—and then listen—at every opportunity: What courses can we offer that are *worthy* of Penn State students?

Folks were interested. The question touched our common concerns. They wanted to talk as well as listen to the ideas of others.

To that end, I asked members of the review committee these questions: (1) What courses will most benefit our students? (2) How can those courses be worthy of our students? (3) What are new, creative, newsworthy ways to rethink such courses? Fortunately, my questions—posed within a context of a faculty-wide uproar—slowed the velocity of the review at the same time that they promoted the development of updated criteria for our required writing and speaking courses.

We now have new criteria because, rather than depending on a committee to remodel our writing and speaking courses, a small committee of us (all feminist rhetoricians) wrested the responsibility from the faculty senate and developed readily approved criteria ourselves.[23] These criteria have nothing to do with themes—and everything to do with rhetorical capaciousness.[24] After all, as Micciche and many others continue to remind us, we WPAs are the experts. All it takes to remind others of our expertise is to employ our "knowledge of the history and current status of writing instruction" in order to "devise curricula, conduct teacher-training courses and faculty development workshops, establish goals for writing programs, and determine assessment procedures" (440). Given our collective expertise, then, we five feminist rhetoricians developed frameworks for the three university-wide required writing and speaking courses, criteria that meet the Council of Writing Program Administrators' Outcomes Statement as well as the Statement of Principles for Public Speaking as a Liberal Art.

As we move into the future, then, the following four criteria will serve as the foundations for Penn State's required writing and speaking courses, criteria anchored in Aristotle's definition of rhetoric and in James Herrick's gloss (from his chapter in this collection) of the definition: (1) Students will develop rhetorical knowledge;[25] (2) students will develop capacities in critical thinking, reading, listening, and the generation of ideas;[26] (3) students will develop proficiency in composing processes;[27] and (4) students will develop knowledge of communication conventions.[28]

Other than the very good news that rhetoric, writing, and speech faculty from all across our campuses work well together is the maybe even better news that members of the faculty senate now realize that our writing and speaking courses actually already had content and that it was never incumbent on them to supply content (in the guise of themes) for our courses. My colleagues and I continue to work together to envision even more engaging, transdisciplinary writing, speaking, and communication courses (whether upper division or lower division, resident instruction or online) that are always/already anchored in rhetorical principles. Together we can imagine an academic future of courses that both nourish and support student writers, speakers, and composers as they draft, write, and revise the kinds of communications that they value and we all like to read and hear. As our department and the PWR move forward into the future, we will face many more curricular—and I would add—hiring and staffing challenges qua opportunities. It is my hope that my feminist rhetorical principles will continue to sustain and energize me.

Coda

We're not trapped in the present. . . . We do have choices. We're living through a certain point of history that needs us to live it and make it and write it.

—Adrienne Rich, "Arts"

In *Feminist Rhetorical Practices*, Gesa Kirsch and Jacqueline Jones Royster lay out four practices "for articulating how researchers and scholars are stretching the boundaries of our work to reach beyond the basics . . . toward the development of new paradigms for how our work itself might be shaped" and deployed (13–14). Those practices include (1) critical imagination, (2) strategic contemplation, (3) social circulation, and (4) globalization (19–25). Such practices are widely employed by feminist rhetorical WPAs, all of whom are working hard to live, make, and write their programs into the future.

In terms of "critical imagination," for instance, many of us, even those of us who regularly publish in rhetoric and/or writing studies, still consider our administrative work to be part and parcel of our work. After all, as Charles Schuster assures us, WPAs "must possess both administrative skills and broad-based, up-to-date knowledge of highly specialized theory and practice" (ix). We tap our disciplinary expertise to work through issues such as a general education review when we reach out to collaborate, listen to, and understand our colleagues, all the while finding ethical ways to reenvision the status quo or to transform an initially "bad" idea into a much greater good. Maybe not the "greatest good," signified by *eudaimonia*, but a greater good, to be sure.

When feminist rhetorical WPAs leverage critical imagination, we clear a space for "strategic contemplations" that allow us to remember, even imagine, what it is that we know, who we know, how we know, and what works. Productive silence, respectful listening, and meditation—these are elements of strategic contemplation that fuel our eventual risk taking, our moving through conflict toward resolution. Whether a colleague "goes rogue" in an instructor-preparation course, a new lecturer reveals that her repeated absences are related to her bipolar disorder, a first-year student reports that he has "nothing to learn" in class, or a harassed graduate student breaks down in our office, feminist rhetorical WPAs give everyone—especially themselves—the space to think and talk toward possible solutions. A feminist rhetorical WPA can quietly confront the rogue colleague, talking through the behavior and imagining together some possible solutions. She can invite the lecturer for coffee to find out the reasons behind the absences, and she can meet with the first-year student to find out what is actually going on in his life. Strategic contemplations are not mystical; they are authentic encounters of creatively working together toward that greater good.

The "social circulation" element of feminist rhetorical practices allows for the contextualization of any event within a scene, a Burkean scene with acts, agents, agency, and purpose. The "politics of [our] location" comes into consideration as well as our own personal, professional, and public histories, which often serve as reliable sources of knowledge. Those of us who find purchase in feminist histories and the feminist present have only to invoke Hillary Clinton, Ruth Bader Ginsberg, Audre Lorde, Adrienne Rich, Sonia Sotomayor, Virginia Woolf, Malala Yousafzai, and many, many others (including Wilma Ebbitt) to envision how strong, intelligent, purposeful women have judiciously and productively negotiated their stances, their plans for the future. At this point in our feminist rhetorical practices, consideration of women's failures is every bit as important as our successes, whether we are facing the challenges of helicopter parents, aggressive students, imbecilic administrators, disrespectful colleagues—or the opportunities to evolve, build, grow, and work even smarter toward our goal.[29] Feminist rhetorical WPAs can invoke feminist legacies of action and performance to model new scenes of writing, teaching, and administering, just as the five of us did when we rewrote the new criteria for our required speaking and writing courses.

The "globalization" feature of Kirsch and Royster's practices may be the most futuristic of their features. As waves of international students meet on our shores, enroll in our courses, and eventually move through our universities as graduate students and professors and through our communities as the professionals they came here to become, all of us are forced to pay especial attention to their cultures, languages, and ways of being. Their interests in our culture have prompted many feminist rhetorical scholars and WPAs to reciprocate such interest. No longer does any feminist rhetorical history or theory seminar stop at our national boundaries. Scholars have much to learn from women like Aung San Suu Kyi, Benazir Bhutto, Angela Merkel, and the iron girls of post-Mao China (Wu), women who have worked or continue to work toward positive change, toward *eudaimonia*. Together, these feminist rhetorical practices provide us all—masculinists, feminists, combinations, or in-betweens—new ways to map our locations, to chart our course to new places, even to imagine the new places we want to inhabit.

Am I a perfect feminist rhetorical WPA? Not even close. Do I continue to pursue my goals as such? You bet. Do I like the work? I do. Do I like preparing teachers, developing the curriculum, working with instructors and graduate students, hearing out frustrated students? I do. Would I have enjoyed this work earlier in my career? Probably not. Would I enjoy the work in another location, within another mesosystem? That is a good question—I do not know, for a WPA position at another university, with other colleagues, comprises a different ecology with different pressures, needs, expectations, and opportunities.

Am I successful? Often, but not always, if success is based solely on the masculinist, competitive, winner-take-all model that we have too long employed. As a feminist rhetorical WPA, however, I can consider my so-called successes and failures within a rich and complex system that includes other people, other programs, other values. Thus, every time I am successful, I give credit to the strong women who have gone before. When I am not successful, I consider how I could have employed my rhetorical skills better; I rethink the rhetorical situation to consider whether there was actually any chance for change in the first place. When I fail, I try hard to convert that failure into needed information for the future. And when I consider the material conditions of my location, I am grateful to Fred Pattee, who established the prominence of rhetoric early on; to Wilma Ebbitt, who set the course for a rhetorically based writing program; and to my hardworking colleagues, who are willing to talk with me about student learning, transfer, assessment, grading, discourses, assignment design, adjunct labor, and textbooks. Male and female alike, they advise, help, reach out, collaborate, and listen with me, often within a feminist rhetorical framework. And so it is with them, as well as with feminist rhetoricians, that my "lot is cast" and whence my "strength come[s]" (Rich, "IV" 6).

Feminist rhetorical practices help me translate growth, improvements, and setbacks within an alternative system of success. These practices have habituated me to working toward the development of an ideal project (a writing program, for instance) at the same time that I nurture a capacity for mutual respect, give-and-take, negotiation, and, especially, for understanding the limitations necessary for the good of the community, that greatest good, *eudaimonia.* Resources, recognition, and hiring do not necessarily make for a zero-sum game when one is working in the ecosystem that is a comprehensive university. In other words, feminist rhetorical practices help me put limitations and resources into perspective, help me understand that getting my own way, pushing my own argument, may not always be the best action or result. Negotiations and counterarguments might well result in an idea better than my original argument; they might result in turn taking; they might even result in friendships. Surely, they result in invitations to return to the negotiating or planning table—whether I am working with members of the faculty senate, faculty from across the university, or my own PWR colleagues.

You never know. Some of the people I work with may be feminist rhetoricians as well.

I dwell in Possibility –   —Emily Dickinson, "466"

Notes

1. Although surrounded by self-identified feminists such as Adrienne Rich, Barbara Christian, June Jordan, Toni Cade Bambara, and her good friend Janet Emig, Shaughnessy was not—to my knowledge—a feminist. Those who worked with her on special projects during the 1970s, Andrea Lunsford and Dixie Goswami, for instance, noticed no connection between Shaughnessy and feminism. As Goswami writes, "I know of no writing or advocacy actions in which Mina expressed or presented herself as a feminist," and she advises me to focus on Emig, "this towering figure," instead. Shaughnessy biographer Jane Maher confirms that "Janet Emig has described Mina as 'breathtakingly nonfeminist,'" continuing with Rich's remembrance of Shaughnessy as "good-natured" about feminism, a movement she regarded as having no "possible relevancy . . . to her life and concerns" (144).

2. These women employed their academic backgrounds and critical imaginations in the service of preparing a curriculum for students who had neither the grades nor the money to attend college. At the City University of New York, Kenneth Bruffee was doing the same, as he was developing long-lasting practices of peer tutoring and collaborative learning.

3. However, in *Yearning*, bell hooks invokes such margins as "spaces for radical openness" (145).

4. My sense is that *coordinator* is the term used most often, given that it registers so little power without diminishing the responsibility. Maybe it was just that thinking that motivated me to change my own title, a move I explain in note 11.

5. Another questionable choice of nomenclature, as though students and their sentences need medical assistance.

6. Echoing Holbrook, Enos reminds us that doctoral programs tend to call themselves "rhetoric and writing" as a way to distinguish themselves from "composition," first-year service work. She goes on to say that "rhetoric is recognized as more of a discipline while composition is simply a service-oriented field with no accepted theory or methodology" (79).

7. Thus, Bloom carefully points out the perceived difference between "work" (that is, service) and "scholarly work":

> I want a Writing Director who will not demand attention when I am pre-occupied with my scholarly work, and who will remain faithful to my needs so that I do not have to clutter up my intellectual life with administrative details. And I want a Writing Director who understands that *my* professional work may involve neglecting her in order to relate to my literary critic colleagues as fully as possible. (177)

8. The authors replicated Linda Peterson's 1986 survey of Council of Writing Program Administrators, which was titled "The WPA's Progress."

9. In *Yearning*, hooks translates the "politics of location" into a stance of re-vision, a realm of counterhegemony, a movement "out of one's place," a confrontation of choice and location (145).

10. One of the first reconsiderations I had was to change the name of the program from the Composition Program to the Program in Writing and Rhetoric. I also felt it was important to change the administrative title from *coordinator* to *director*. These two proposed changes were put to a vote, and the great majority of my colleagues voted in favor of them. (I still wonder why two people voted no.)

11. My location is University Park, the main campus of Penn State. Many of the other Penn State campus have impressive writing and rhetoric faculty, including Laurie Grobman, Kate Latterell, Danielle Mitchell, and Christian Weisser, to name just a few.

12. Throughout, I work to distinguish among the graduate students, lecturers, and faculty who constitute the cohort of writing instructors.

13. I am phasing out our noncredit developmental writing course and replacing it with a full-credit hybrid-studio course. The hybrid-studio version of first-year writing includes a smaller class size, the regular curriculum, weekly tutorials with a graduate student, and assignment-specific peer-to-peer tutorials in the writing center. The course has been so successful that many of the other Penn State campuses are using the same model, just as they are using our model of guided self-placement into our first-year courses (developmental, regular, and honors).

14. The two-course sequence called Rhetoric and Civic Life constitutes a collaboration between the rhetoric faculty in communication arts and sciences and in English, with graduate students and lecturers from both departments teaching in the program. Since its 2009 rollout, the multimodal, rhetorically rich sequence has been a success.

15. We offer upper-division rhetoric courses in history, theory, and translingualism as well as courses in cross-cultural, digital, collaborative, feminist, cultural-ethnic, and writing-center-focused rhetorics as well.

16. The administrative staff in English believes that there are already enough sections of business writing to accommodate the Smeal students but that those students are not willing to take sections of business writing unless they are offered on Tuesdays and Thursdays between ten o'clock and two o'clock.

17. I like to think that I employed *metis*, which, as Hawhee explains, is a cunning intelligence that is employed to negotiate agonistic forces. She explains that *metis* exists only when it is practiced (11, 47, 53).

18. My department head, my associate dean, and I negotiated with the dean of business, dean of liberal arts, and the provost to underwrite these three positions for

two years (thinking that we could widen the bottleneck in those two years); I had argued for hiring only two assistant professors for three years, on the grounds that we could not do so in so short a time.

19. Although the teaching load is the same for these new hires as for the VAPs, the salaries are lower—but the terms are longer. My department head and I worked closely together to write the job ads. I interviewed candidates with the help of the associate head and an experienced lecturer. All of these collaborations could be considered feminist, as they were all strategic, respectful, collaborative, and, ultimately, productive.

20. In other words, the number of students registered for our courses determines the number of sections we will offer and, therefore, the number of writing instructors we can hire.

21. Our dean may or may not decide to fund tenure lines, regardless of departmental votes and priority rankings.

22. In fact, a good many of us in the English department wrote responses to the proposed general education changes, reminding members of this faculty senate committee that "themes and clusters" were much like "majors and minors." We also spoke out against the proposed themes for first-year writing, which (as rumor had it) included the themes of "life and death," "love and sex," and "a continent." I took the concept of a continent and ran with it, taking the lead on the controversy, as H. Miller would have a feminist WPA do, on occasion. On the surface, the theme was so ridiculous that it served as an easy target, even when members of the committee replied that it was "just one idea."

23. Then associate dean Christopher Long was leading part of the general education review and, by some fortunate sequence of events, began collaborating on the writing and speaking requirements. When he asked me whom I would like to work with, I quickly named leaders in rhetoric, writing, and speech communication at University Park and other Penn State campuses, including Laurie Grobman, Michele Kennerly, Mary Miles, Danielle Mitchell, and Molly Meijer Wertheimer. These feminist rhetoricians joined our group, and we managed to dispatch a set of disciplinary-based criteria in two days. Associate Dean Long was impressed.

24. My so-far-unsuccessful argument is to offer more (not fewer) writing and speaking requirements. After all, there's every amount of research in the world to assure any university (especially a leading one like Penn State) that a one-shot writing inoculation (first-year writing alone) does very little to help students improve their writing over the long term.

25. *Rhetorical knowledge* is the ability to analyze contexts and audiences and then to act on that analysis by determining appropriate lines of inquiry and creating a range of written and oral responses. Writers and speakers develop and apply rhetorical knowledge by negotiating purpose, audience, context, genre, and conventions as they explore, compose, and deliver a variety of texts for ever-evolving situations and media.

26. *Critical thinking* is the ability to analyze, synthesize, interpret, and evaluate ideas, information, situations, and texts. When writers and speakers think critically about the materials they use—whether print texts, photographs, data sets, videos, speeches, or other materials—they separate assertion from evidence, evaluate both

sources and supporting evidence, recognize and evaluate underlying assumptions, read across texts for connections and patterns, identify and evaluate chains of reasoning, and compose appropriately qualified and developed claims and generalizations. These intellectual practices serve as the foundation of advanced academic writing and speaking.

27. Writers and speakers use multiple strategies, or *composing processes*, to conceptualize, develop, and finalize projects. Composing processes are seldom linear: a writer or speaker may research a topic before drafting, then conduct additional research while revising or after consulting a colleague. Composing processes are also flexible: successful writers and speakers can adapt their composing processes to different contexts, audiences, and occasions.

28. *Conventions* are the formal rules and informal guidelines that define genres and, in so doing, shape perceptions of correctness or appropriateness. Most obviously, conventions govern such things as mechanics, usage, spelling, pronunciation, and citation practices. But they also influence features of content, style, arrangement/organization, graphics, delivery (or presentation), and design. Conventions are not universal, however; they vary by genre (conventions for lab notebooks and discussion-board exchanges differ); by discipline (conventional moves in literature reviews in psychology differ from those in English); and by occasion (meeting minutes and executive summaries, extemporaneous remarks and official addresses use different registers). A writer's or speaker's grasp of conventions in one context does not equate to a firm grasp in another. Successful writers and speakers understand, analyze, and negotiate the conventions associated with different purposes, audiences, contexts, and genres as they analyze, compose, and deliver messages in a range of written, oral, and digital media.

29. Clinton's "smart power" comprises a set of feminist rhetorical practices that initiate and maintain alliances for establishing "long-lasting trends for a better tomorrow" ("Smart Power"). Specific practices include reaching out and working through, productive silence and respectful listening, collaboration, and judicious speaking.

Works Cited

Alcoff, Linda. "The Problem of Speaking for Others." *Who Can Speak? Authority and Critical Identity*, edited by Judith Roof and Robyn Wiegman, U of Illinois P, 1995, pp. 92–119.

Aristotle. *The Rhetoric and the Poetics*. Translated by W. Rhys Roberts, Modern Library, 1984.

Barr-Ebest, Sally. "Gender Differences in Writing Program Administration." *WPA: Writing Program Administration*, vol. 18, no. 3, 1995, pp. 53–73.

Bhutto, Benazir. *Daughter of Destiny: An Autobiography*. Harper, 2005.

Bloom, Lynn Z. "I Want a Writing Director." *College Composition and Communication*, vol. 43, no. 2, 1992, pp. 176–78.

Burke, Kenneth. *A Rhetoric of Motives*. U of California P, 1969.

Charlton, Jonnika, and Shirley K. Rose. "Twenty More Years in the WPA's Progress." *WPA: Writing Program Administration*, vol. 33, nos. 1–2, 2009, pp. 114–45.

Clinton, Hillary Rodham. *Hard Choices*. Simon and Schuster, 2014.

———. "Smart Power: Security through Inclusive Leadership." Georgetown U, Washington, DC, 3 Dec. 2014. Lecture.

Dickson, Marcia. "Directing without Power: Adventures in Constructing a Model of Feminist Writing Program Administration." *Writing Ourselves into the Story: Unheard Voices from Composition Studies*, edited by Sheryl I. Fontaine and Susan Hunter, Southern Illinois UP, 1993, pp. 140–53.

Enos, Theresa. *Gender Roles and Faculty Lives in Rhetoric and Composition*. Southern Illinois UP, 1996.

Foss, Sonja, and Karen Foss. *Inviting Transformation*. 3rd ed., Waveland, 2011.

Glenn, Cheryl. *Rhetoric Retold: Regendering the Tradition from Antiquity through the Renaissance*. Southern Illinois UP, 1997.

———. *Unspoken: A Rhetoric of Silence*. Southern Illinois UP, 2004.

Glenn, Cheryl, and Krista Ratcliffe, editors. *Silence and Listening as Rhetorical Arts*. Southern Illinois UP, 2011.

Goodburn, Amy, and Carrie Shively Leverenz. "Feminist Writing Program Administration: Resisting the Bureaucrat Within." *Feminism and Composition Studies: In Other Words*, edited by Susan C. Jarratt and Lynn Worsham, Modern Language Association, 1998, pp. 276–90.

Goswami, Dixie. "Feminism!" Email message to the author, 7 May 2015.

Gunner, Jeanne. "Decentering the WPA." *WPA: Writing Program Administration*, vol. 18, nos. 1–2, 1994, pp. 8–15.

Hawhee, Debra. *Bodily Arts: Rhetoric and Athletics in Ancient Greece*. U of Texas P, 2004.

Holbrook, Sue Ellen. "Women's Work: The Feminization of Composition." *Rhetoric Review*, vol. 9, no. 2, 1991, pp. 201–29.

hooks, bell. *Yearning: Race, Gender, and Cultural Politics*. South End Press, 1990.

Kyi, Aung San Suu. *Freedom from Fear*. Penguin, 2010.

Lipari, Lisbeth. *Listening, Thinking, Being: Toward an Ethics of Attunement*. Penn State UP, 2014.

Logan, Shirley Wilson. *With Pen and Voice: Critical Anthology of Nineteenth-Century African-American Women*. Southern Illinois UP, 1995.

Lorde, Audre. "Learning from the 60s." 1982. *Sister Outsider: Essays and Speeches by Audre Lorde*. Crossing Press, 2007, pp. 134–44.

Lunsford, Andrea A. *Reclaiming Rhetorica*. U of Pittsburgh P, 1995.

Maher, Jane. *Mina P. Shaughnessy: Her Life and Work*. National Council of Teachers of English, 1997.

Micciche, Laura R. "More than a Feeling: Disappointment and WPA Work." *WPA: Writing Program Administration*, vol. 64, no. 4, 2002, pp. 432–58.

Miller, Hildy. "Postmasculinist Directions in Writing Program Administration." *WPA: Writing Program Administration*, vol. 20, nos. 1–2, 1996, pp. 49–61.

Miller, Susan P. *Textual Carnivals: The Politics of Composition*. Southern Illinois UP, 1991.

Peterson, Linda. "The WPA's Progress: A Survey, Story, and Commentary on the Career Patterns of Writing Program Administrators." *WPA: Writing Program Administration*, vol. 10, no. 3, 1987, pp. 11–18.

Ratcliffe, Krista. *Rhetorical Listening: Identification, Gender, Whiteness.* Southern Illinois UP, 2005.

Ratcliffe, Krista, and Rebecca Rickly, editors. *Performing Feminism and Administration in Rhetoric and Composition Studies.* Hampton Press, 2010.

Rich, Adrienne. "IV." *Your Native Land, Your Life.* Norton, 1986, p. 25.

———. "Toward a Woman-Centered University." *On Lies, Secrets, and Silence: Selected Prose, 1966–1978.* Norton, 1979, pp. 125–55.

Ronald, Kate, Cristy Beemer, and Lisa Shaver. "'Where Else Should Feminist Rhetoricians Be?' Leading a WAC Initiative in a School of Business." Ratcliffe and Rickly, pp. 159–69.

Royster, Jacqueline Jones. *Traces of a Stream: Literacy and Social Change among African American Women.* U of Pittsburgh P, 2000.

Royster, Jacqueline Jones, and Gesa Kirsch. *Feminist Rhetorical Practices: New Horizons for Rhetoric, Composition, and Literacy Studies.* Southern Illinois UP, 2012.

Schell, Eileen. *Gypsy Academics and Mother-Teachers: Gender, Contingent Labor, and Writing Instruction.* Boynton/Cook, 1998.

Schuster, Charles. Foreword. *Resituating Writing: Constructing and Administering Writing Programs,* edited by Joseph Janangelo and Kristine Hansen, Boynton/Cook, 1995, pp. ix–xiv.

Shaughnessy, Mina P. *Errors and Expectations: A Guide for Basic Writing Teachers.* Oxford UP, 1977.

Sotomayor, Sonia. *My Beloved World.* Knopf, 2013.

Stallybrass, Peter, and Allon White. *The Politics and Poetics of Transgression.* Cornell UP, 1986.

Syfer, Judy. "I Want a Wife." *Ms.,* vol. 1, no. 1, Dec. 1971, p. 13.

Wu, Hui. *Once Iron Girls: Essays on Gender by Post-Mao Chinese Literary Women.* Lexington Books, 2009.

Yancey, Kathleen Blake. "Defining Moments: The Role of Institutional Departure in the Work of a (Feminist) WPA." Ratcliffe and Rickly, pp. 143–58.

Yousafzai, Malala, and Christina Lamb. *I Am Malala.* Little, Brown, 2013.

Section III
Language Perspectives

Language as System and Situation: Writing the World

James Paul Gee

Since language is never divorced from experiences in the world, Gee explains that it is experience that gives both oral and written language situational, contextually appropriate, properly nuanced meanings. Paulo Freire, in *Pedagogy of the Oppressed*, pointed out that reading the world and reading the word are integrally connected. When we situate writing within experiences, the world we enable for students allows them to both read the word and write the world. When detached from experiences in the world, writing can quickly become a sterile exercise in abstraction. The accountability that matters in literacy teaching is accountability to the world of experience and the relationships between language as system (grammar) and language as situational meanings in contexts of taking action in the world.

Language as System

Language is obviously crucial to any discussion about writing. However, there are two important ways to talk about language that should be clearly distinguished (Gee, *Unified Discourse Analysis*; Levinson). The first way we can talk about language is to call a language a *system*. Any language is a grammar of experience. The grammar, or structure of the morphemes, words, phrases, and sentences in a language, is a system that cuts up, regiments, and labels experience in certain ways. The system affords speakers of the language a certain way of looking at the world.

For example, Standard English uses auxiliary verbs and suffixes on verbs to distinguish between actions that are viewed as a completed point in time ("He washed his car") and actions that are viewed as ongoing in time ("He was washing his car"), though, of course, all actions take time to unfold. Varieties of English spoken by some African Americans make a third distinction (Gee, *Social Linguistics*). They use a "bare *be*" form to mean that an action is habitual or recurs regularly over time ("He be washing his car" = he washes his car all the time / he habitually washes his car). This example makes clear that even closely related dialects can map the world in different ways, though, of course, unrelated languages can display more dramatic variations.

It is not just languages and dialects that map the world as system differently, so do different registers or social languages, that is, styles or varieties of language that are associated with particular identities, groups, institutions, or endeavors (Gee, *Social Linguistics*). The language of physics uses the word *work* differently than does vernacular English. The language of chemistry uses words like *heat* and *temperature* differently than does vernacular English. The language of video gamers is different from the language of doctors or gamblers. This is so because gamers, doctors, and gamblers are interested in making different distinctions and dividing up categories differently from one another and from vernacular English.

Consider a difference like "Hornworms sure vary a lot in how well they grow" (in vernacular English) and "Hornworm growth exhibits a significant amount of variation" (in academic language of the sort associated with domains like biology). The former sentence, in the vernacular, asks us to see the world as made up of concrete things (like hornworms) and dynamic processes of change (like growing and varying). The latter academic sentence asks us to see the world as made up of abstract measurable variables like growth and variation. Furthermore, the former sentence is based on one's own observation of hornworms and represents one's opinion. The latter sentence is based on tests of significance owned by an academic field and represents a much more "anonymous" discipline-based claim than the former. Note then that "Hornworm growth sure exhibits a significant amount of variation" sounds odd. The affective/emotive term *sure* does not fit with the "voice of reason and evidence" style of academic language. Each form of language gives us a different stance toward the world of experience. Hornworms come to mean and even, in a sense, be different things.

Language as system is a lot like a set of clothes. The sort of clothes we wear to the beach—what goes with what—is different from the sort of clothes we wear to a business office, and this again is different on a casual Friday. Note that just as each set of clothes is associated with a different identity, function, and context, so, too, are different dialects and styles of language (social languages).

The world can be cut up in different ways for different purposes by groups with different interests. For example, gamers distinguish among role-playing games, action games, adventure games, strategy games, and platformers, among others. Games have many different features, and the world of games could be cut up in many different ways. How we cut them up into categories is determined by the words we have, and the words we have are determined by the efforts and interests of groups to which we belong. We could imagine someone with other interests categorizing games as games with avatars, games without avatars, games with stories, games without stories, games with first- or third-person perspective versus "god games" (with a god's-eye view of the world). These are two different systems, and the latter will only exist if and when some group has things they need to do and people they need to be that would render this system useful and important to them.

Some linguists use the term *utterance-type meaning* for the system-level meanings to which grammar gives rise (Levinson). At the "type" level, *cat* means a feline animal and belongs to sets of related terms like *pet, domesticated animal, small animal,* and so forth. At the level of *utterance-token meaning,* a word like *cat* can take on quite specific and even novel meanings in specific contexts of use. For example, if we are staring at a cloud shaped like a cat and I say, "The cat is moving a way," the word *cat* means a cat-shaped cloud. It is to the issue of utterance-token meanings that I now turn.

Language as Situated

The second important way we can talk about language is as *situated within contexts of use* (Gee, *Unified Discourse Analysis*; Clark, *Microcognition*). At a system level some varieties of English distinguish among things like coffee, tea, soda, juice, water, beer, and wine as drinks. (For example, this is a list I could offer you as a guest when I ask you if you want something to drink.) However, when we actually use these words, they can take on different meanings in different contexts. If I say, "the coffee spilled; go get a mop," I mean coffee as liquid; if I say, "the coffee spilled; go get a broom," I mean coffee as grounds; if I say, "the coffee spilled; stack it again," I mean coffee as tins; and if I say, "the coffee spilled; scoop it up," I mean coffee as coffee-flavored ice cream. In fact I can say something as novel as "Big Coffee is as bad as Big Oil," and you know how to situate the meaning of "Big Coffee" based on all the contexts in which you have heard or uses "Big Oil." If I were asking volunteers at a food bank if they wanted me to donate coffee, tea, soda, juice, water, beer, or wine, the list of terms would no longer name drinks but rather cans and bottles, and those "in the know" would know that adding beer and wine is probably a joke in this context.

Consider that a sentence like "Any society with thoroughgoing restrictions on economic freedom could not be a democracy" at the system level seems to be nonsense. *Democracy* at the system level just means voting or voting for representatives who vote. What could possibly stop any society from voting for any economic restrictions it wished? Yet this sentence expresses a central claim of the sort of neoliberal economics championed by the late Milton Friedman. You can only assign a situated meaning to *democracy* here if you know Friedman's theories and those of his followers (Gee, *Social Linguistics*). Once you know this context, the use of the word is no longer nonsense, though the theory may, of course, be false or true.

Notice that the distinction between system and situation is not one of acultural versus cultural. Some social groups of English speakers use "coffee, tea, juice, soda, water, beer, or wine" as a systematic set of "common drinks" and of options to offer guests as drinks on social occasions. Some groups have other sets of terms and drinks. Some social groups are interested in distinguishing films and video games by appropriate ages for play and by content that is violent or nonviolent, sexually suggestive or not, and other such distinctions. These are the categories they have. Gamers are not all that interested in such categories (unless they are playing their parent role) but tend to be interested in a different set of categories. Both systems and situations can vary by social group and culture.

A Chicken and Egg Paradox

Which comes first, system or situation? We only know what *coffee* means fully when we have experienced a range of situated meanings the word can have, and we can even make up new situated meanings for it as contexts or cultures change. But we can only situate the meanings of words we have in our language (or any given social language) and in terms of how these words relate to the other words and to the syntax of that language or social language.

In turning to structures, we consider more than just words. While the subjects of sentences in English are always topic-like (this is their general system meaning), in different situations of use, subjects take on a range of more specific meanings. If, in a debate, I say, "The constitution only protects the rich," the subject of the sentence (constitution) is an entity about which a claim is being made; if a friend of yours has just arrived and I usher her in saying, "Mary's here," the subject of the sentence (Mary) is a center of interest or attention; and in a situation where I am commiserating with a friend and say something like "You really got cheated by that guy," the subject of the sentence (you) is a center of empathy (signaled also by the fact that the normal subject of the active-voice version of the sentence—"That guy really cheated you"—has been "demoted" from the subject position through the so-called get-passive construction). Which comes first, subjects as topics (their meaning or function in the system) or the range of situated meanings topics can have in contexts of use?

Clothes work the same as language here again. Wearing a black suit as a system of coordinated shirt, pants, jacket, tie, and shoes (all of which must be black or go with black) always connotes "formal." But it means a different thing at a business office than it does at a funeral than it does when worn by a college teacher (who is either of an earlier generation or teaches something like business) than it does when worn by a minister (who is, thus, not a Catholic priest). Which comes first, a black suit with the system of coordinating and contrasting sorts of clothes into which it fits or the range of meanings a black suit can (and cannot) have?

Language as system regiments and categorizes reality and renders it recognizable (Vygotsky). Language as situation uses experiences in the world to give language a range of meanings without which, of course, there would be no meanings at all. Language gets meaning from experience (it has to mean something when it is used), and yet it organizes the very experience that gives it meaning. Which comes first? Neither. They bootstrap each other into existence.

Imagine that a whole new genre of video game is invented. We could now apply one of our current system's labels to this new genre and, thus, widen the range of situated meanings that existing label can have. Or we could create a new label or a new subcategory of an old label and thereby change the system, since now all the old labels have a new label to comport within the system. In these ways language as both system and situation evolves.

The same is true again of clothes: those socks with toes are called "toe socks" (a new subcategory), though we could have called them "foot gloves." On the other hand, for some people, given the modern debates around abortion, with the new abilities of medicine to keep premature babies alive, the word *human* has been given a new situated meaning in terms of which even a very early fetus is now a "human" and "life" begins at conception. (But note that no one, I think, would say a fertilized egg is a chicken.) This latter example makes it clear that how system relates to situation and how they evolve as things change can be quite consequential, even to the heart and soul of what constitutes civil society and creates or resolves cultural divides.

Language, Experience, and Learning

So we have concluded that language gets meaning (situated meaning) from experience and that experience gets meaning (system meaning, by getting categorized and regimented) from language. So what are the implications of all this for teaching and learning? The implications start here: Humans learn from experience. It used to be believed that the human mind works much like a digital computer (Pylyshyn). Humans, like computers, were thought to learn and think through calculation, abstraction, symbols, and generalities. But in fact humans tend to be bad at what digital computers are good at and good at what they are bad at—that's why we have them. Computers are good at keeping budgets, bad at facial recognition. Humans are often poor at balancing their checkbooks but unbelievably good at recognizing faces. A newer theory of mind—based on situated or embodied cognition—argues that humans are pattern recognizers par excellence (Barsalou, "Language Comprehension" and "Perceptions"; Bergen; Glenberg; Glenberg and Robertson; Glenberg and Gallese). Humans have experiences in the world; they store all these in their long-term memories (which are, unlike a digital computer, nearly limitless); and they use these stored memories to search for patterns and subpatterns that will help them prepare for action in new experiences.

None of this means that humans cannot abstract and generalize. What it means is that humans often do so from the bottom up (diSessa). They start with experience and with concrete understandings, and as they gain more experience, they find ever more generalizable patterns and eventually gain more general knowledge. Humans cannot learn well from texts until they have experienced how the sorts of words and grammar of the social language in texts translate into experience—that is, until they can associate the words of the texts with images, actions, goals, and dialogue (which are situated meanings). Once they have experienced situated meanings in a variety of relevant contexts, they are ready to learn by reading and hearing larger blocks of language out of context—but only then.

Humans can use their stored experiences to run role-playing simulations in their minds to prepare for action (Gee and Hayes, *Language*). For example, if you want to prepare for a toast you have to give at a friend's wedding, you can role-play various scenarios—even role-play the reactions of different people at the event—to make a decision about what to do and what not to. To do this you will use experiences you have had, but also ones you have heard about or seen, for example, in movies. (Humans do not strongly distinguish between experiences they have had in the world and the ones they have had via media, according to Reeves and Nass).

But experiences are best for human learning when they have certain key features. Let me define a new term, experience^{GL}, meaning "experience that is good for learning." An experience^{GL} has the following features (Clark,

Being There; Gee, *Situated Language* and *Good Video Games*; Glenberg, "What"; Wason, "Reasoning" and "Reasoning about a Rule"):

1. A learner has an action to take and a clear goal.

2. The learner cares about the outcome of the action—something is at stake for the learner in the outcome of an action.

3. A learner knows how to appreciate (judge, assess) the outcomes of various attempts to succeed at his or her goal and knows how to go on (what sort of thing to try next) after a failed attempt.

4. The learner can manage his or her attention in the experience—know what to pay attention to, what to focus on, and what to background—so as to avoid cognitive overload (always a danger in experience and in multimodal media, as well, where there are a wealth of things that could be attended to).

Now here is where teachers and teaching come into the picture (Hattie and Yates). Beginners cannot do these things for themselves. They need help formulating goals (and knowing when in a course of action to rethink or reformulate them). They need help judging what are good or promising outcomes, and what are not, in their trajectory of attempts toward a goal. They need help knowing how to go on, help with what are good choices about what to do next. They need help managing their attentional economy, help with where and how to pay attention, what to focus on, what elements to foreground and background in the experience.

When someone in the role of teacher (parent, school teacher, peer, media, or technology) offers these sorts of help, they have designed the experience for the learner. They have created a well-designed experience for learning. All learners need well-designed experiences for learning. Beginners need teachers, and deliberate learners are people who have learned to teach themselves and design their own experiences.

Good Teaching

Good teaching is a form of design, creating well-designed experiences for learning. But what has this got to do with our earlier discussion of language as system and language as situated? Let me explicate this through an example. Consider the following quote: "yet I believe [Milton] Friedman is right that thoroughgoing restrictions on economic freedom would turn out to be inconsistent with democracy" (Becker). If you do not understand this quote, how would I teach you to understand it? Well, you would have to understand the sorts of economic language that the Chicago school of economics (founded by Friedman) used and many neoliberal economists still use today; you would have to understand their words and the ways in which their words call forth sets of categories. You would have to understand the way they recruit grammatical resources to order sentences and texts and, thereby, order experience (what counts as foregrounded and backgrounded, what is asserted and assumed, what is easily sayable and what is not). But how can you understand this system if you have had no experience with how these people talk and write, act and value in the world? But, then, how can you even make sense of this experience if you do not already know how these people divide up experience in terms of their system?

This is our chicken and egg paradox again. We have to learn a system through well-designed experiences in the world, and we cannot make sense of experience without the very system we are trying to learn. To show how good teaching can defeat this paradox, I need now to turn to an example that will not require me to bore you with neoliberal economic theories.

An Example of Good Teaching

Yu-Gi-Oh! is a card came on the order of *Magic: the Gathering* or *Hearthstone*. *Yu-Gi-Oh!* can be played face to face or virtually as a video game. The game has thousands of cards. From these, players select forty cards to make decks with cards that allow certain sorts of tactics and strategies in the game. *Yu-Gi-Oh!* is played by those as young as seven years or as old as thirty or more. Each card contains a picture and print that tells a player what the card is and what the character on the card enables the player to do in the game.

The language on the cards is quite technical and complex. Here is a typical example:

Cyber Raider

Card-Type: Effect Monster

Attribute: Dark | **Level**: 4

Type: Machine

ATK: 1400 | **DEF**: 1000

Description: When this card is Summoned: Activate 1 of these effects.

Target 1 Equip Card equipped to a monster on the field; destroy that target.

Target 1 Equip Card equipped to a monster on the field; equip that target to this card.

Players can turn to rule books on the internet and fan sites to check on disputes about the rules or the proper interpretation of the words on a card. Here is an example of the sort of thing they will see:

> In order to Synchro Summon a Synchro Monster, you need 1 Tuner (look for "Tuner" next to its Type). The Tuner Monster and other face-up monsters you use for the Synchro Summon are called Synchro Material Monsters. The sum of their Levels is the Level of Synchro Monster you can Summon. ("About *Yu-Gi-Oh!*")

So *Yu-Gi-Oh!* uses a particular social language. The examples above are language as system with a vengeance for you if you have never played the game. Without experiences of how things work in game play, you do not know what meanings to assign to these words; but without knowing what the words mean in the *Yu-Gi-Oh!* social language, you do not know what to pay attention to in the experience. So how does *Yu-Gi-Oh!* resolve this paradox through good teaching? (And its makers must do so or else go broke, since they do not run a school but a business, and a massively profitable one at that.)

Yu-Gi-Oh! creates what I will call a Situated Learning System for teaching (Gee, *Situated Learning*; Gee and Hayes, *Language*). The system has the following properties:

1. Game that gives language situated meaning

In the game, players use movements and dialogue to place cards out in front of them in ways that act out the meanings of the term. Players come to associate the terms of *Yu-Gi-Oh!* language with images, actions, dialogues, goals, tactics, and strategies.

2. Rule book and websites that give language system meaning

At the same time, rule books and websites exist as repositories of the *Yu-Gi-Oh!* language as a social language and as language as system. Players can consult these texts as they are ready and need to.

3. Affinity spaces creating diverse learning

Yu-Gi-Oh!—like many other popular-culture activities—is associated with a myriad of interest-driven fan sites ("affinity spaces," in the terms of Gee and Hayes, *Women*) where players of all ages, abilities, and social groups come together in their passion for *Yu-Gi-Oh!* to mentor and resource each other via a wide variety of activities ranging from didactic tutorials to models of good play and coaching of all sorts.

4. Convergent media

The *Yu-Gi-Oh!* franchise puts out books, video games, television shows, movies, and websites that show dramatic stories of *Yu-Gi-Oh!* characters (the characters on the cards) acting out the moves in the game, moves that are described in the *Yu-Gi-Oh!* language on the cards and in the rule books.

5. Attentional economy

Video game versions of *Yu-Gi-Oh!*, in their early, more tutorial-like levels or in play with other players who are acting as teachers, actively help newcomers know what to pay attention to and how to manage their attention effectively. They often do this by simplifying the game at first so that newcomers have few variables—but important ones—to pay attention to.

6. Low cost of failure

Like many video games, the cost of failure is relatively low in *Yu-Gi-Oh!* If you lose, you play again. Game sites can help match you with players at your level, and video game versions of the game match you with other comparable players who may be real people or computer-run players. A low cost for failure encourages exploring, risk taking, and trying new things. It teaches players to view failure not as a judgment but as a tool for learning.

7. Managing complexity

Video games—including *Yu-Gi-Oh!* in its video game forms—are designed in terms of levels, which introduce and manage complexity for learners. Games introduce problems in a generative order whereby earlier problems lead learners to hypotheses and solutions that work well later when players use them with other tools to solve hard problems.

8. Language just-in-time and on-demand

Language as system is given "just in time," that is, in small blocks that a learner can see in application, or "on demand," that is, in larger blocks of information out of context when the learner is ready for them, asks for them, and can make good use of them (as a resource). Note that each card has a small amount of just-in-time language that a player uses and sees in application immediately in the game. The rule books and websites contain language best used on-demand and as part of interactions and discussions outside the game.

9. Time not used as a measure of learning

Like most digitally inspired learning out of school and connected to interests and passions, *Yu-Gi-Oh!* does not judge progress and measure learning by time spent. If one player takes longer or needs more time for lacking background knowledge or preparation, it makes no difference. What matters is that the player sticks to it, persists past failure, and moves toward mastery. Players may win or lose a game of *Yu-Gi-Oh!*, but no player may claim superiority simply for having learned part of the game faster than another player.

10. Modding and production

Today, video games are part of a larger trend, a "maker movement" in which people young and old can produce and not just consume (Andersen; Hitt). With or without credentials, people can become experts at media production, design, news, citizen science, activism, technological innovation, or any other domain from physics to gardening. Gamers often "mod" games; that is, they use software to modify games or to create entirely new ones. *Yu-Gi-Oh!* players can design cards, write fan fiction or game guides, and design tutorials for others.

11. Assessing growth

Digital media and affinity spaces allow for the collection of copious data. These data can be mined to assess players on multiple variables, and in terms of growth across time and different trajectories toward mastery, in comparison with thousands of other players. Such information can be used by players to improve, to evaluate their standing, or to inform other people or the makers of *Yu-Gi-Oh!* products about how best to resource players for development and engagement. There are no "drop out of the sky" one-off tests with a single score. Furthermore, because play is social, affinity spaces offer models of excellence and allow everyone to share high standards set indigenously by social groups engaged in collective learning and play.

Writing the World

What does this story about *Yu-Gi-Oh!* offer us as a moral? It is not that *Yu-Gi-Oh!* is special. In fact, there are lots of similar, perhaps even better (e.g., *Minecraft*), examples I could have given. The moral is this: think of teaching as designing and resourcing a *learning system* with moving parts, just as the *Yu-Gi-Oh!* franchise and its players do. The sort of learning system I have sketched is very different from many of today's school and college classrooms. However, such systems are flourishing out of school and are giving rise to twenty-first-century skills and forms of collaboration and collective intelligence that are not even on offer in many schools and colleges.

Paulo Freire famously argued that reading the world must precede reading the word. He meant this as a call to reform and liberation. However, his claim is also simply empirically true. The cognitive capacities we use to comprehend and understand things in the world, oral language, and written language are one and the same (Biemiller; Gee, *Literacy*; Stanovich). A person cannot comprehend texts well unless they can also comprehend oral languages and states of affairs in the world well. And I have also argued here that situated meanings—understanding how words and linguistic structures attach to images, actions, goals, and dialogue in the world (or virtual worlds)—are crucial for learning language as system, for understanding how and why language as system divides up the world as it does. For any text—whether a physics text or one about games or gardening—learners must live in and muck around in the world that the language of the text is about, so that when they turn to the text, they have enough situated understanding to begin sorting how the language of the text works as a sort of guide to that world.

One of the first things I learned when, in my midfifties, I began to game is that a game manual is pretty worthless unless and until you have played the game for a while. The game world and your actions as a player in the game are what the manual is about and what ultimately allows players to understand in ways that make sense and inform future learning and actions. In many respects, giving learners a physics textbook without letting them "play the game" of physics (engage in physics as a set of problems to solve, with tools and practices to solve them, and in interactions with others) is a good deal like giving a kid a manual with no game. This a happy child does not make.

So, following Freire and my discussion here, we see that experiences, when well designed for learning, mediate between reading the world and reading the word:

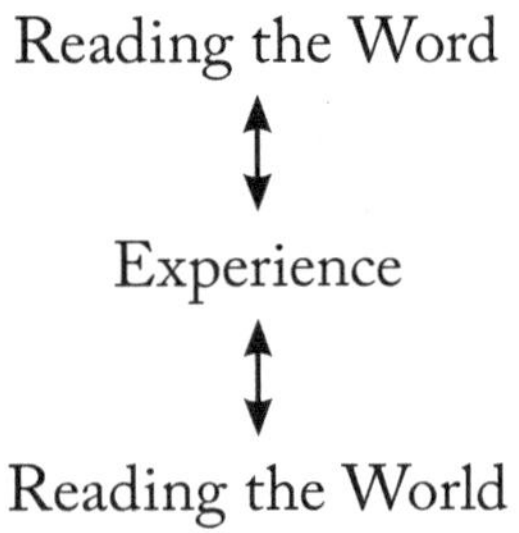

But Freire went even further:

> In a way, however, we can go further and say that reading the word is not preceded merely by reading the world, but by a certain form of writing it or rewriting it, that is, of transforming it by means of conscious, practical work. (Freire and Macedo 35)

So we need another diagrammatic representation:

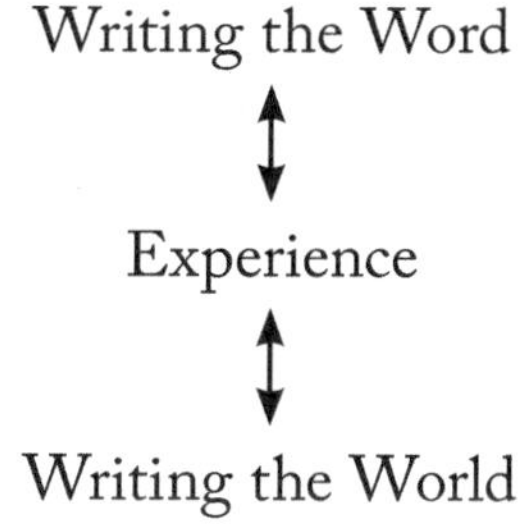

Above, I pointed out that today we live amidst a maker movement. People can use digital tools of all sorts, the internet, and affinity spaces to learn to make almost anything, to become experts without degrees, and to work with others to participate in and produce change without the need for any formal institution. Writing has always been the maker movement part of literacy. While we have long had societies where nearly everyone can read, we have never had societies where writing was anywhere nearly as pervasive.

So how does writing work in a maker movement age, in an age when people can make media, robots, scientific discoveries, art, news, social change, and new knowledge? I would suggest that writing needs to be seen as a maker tool along with other maker tools in the sorts of learning systems with movable parts that I discussed. Such systems are ultimately ways to teach people to play a certain sort of "game" and to join with others to solve problems. In this respect we might see (and teach) writing as ways of making moves in a consequential game in tandem with other tools and other people.

Works Cited

"About Yu-Gi-Oh! Monster Cards." *Yu-Gi-Oh! Card Guide*, www.yugiohcardguide .com/about-yugioh-monster-cards.html.

Andersen, Chris. *Makers: The New Industrial Revolution*. Crown Business, 2012.

Barsalou, Lawrence W. "Language Comprehension: Archival Memory or Preparation for Situated Action?" *Discourse Processes*, vol. 28, 1999, pp. 61–80.

———. "Perceptions of Perceptual Symbols." *Behavioral and Brain Sciences*, vol. 22, no. 4, 1999, pp. 637–60.

Becker, Gary. "On Milton Friedman's Ideas." *The Becker-Posner Blog*, 19 Nov. 2006, www.becker-posner-blog.com/2006/11/on-milton-friedmans-ideas--becker.html.

Bergen, Benjamin K. *Louder than Words: The New Science of How the Mind Makes Meaning*. Basic Books, 2012.

Biemiller, Andrew. "Oral Comprehension Sets the Ceiling on Reading Comprehension." *American Educator*, vol. 27, no. 1, 2003, pp. 23, 44.

Clark, Andy. *Being There: Putting Brain, Body, and World Together Again*. MIT P, 1997.

———. *Microcognition: Philosophy, Cognitive Science, and Parallel Distributed Processing*. MIT P, 1989.

diSessa, Andrea A. *Changing Minds: Computers, Learning, and Literacy*. MIT P, 2000.

Freire, Paulo. *Pedagogy of the Oppressed*. 1968. Translated by Myra Bergman Ramos, Continuum, 1995.

Freire, Paulo, and Donaldo Macedo. *Literacy: Reading the Word and the World*. Bergin and Garvey, 1987.

Gee, James Paul. *Good Video Games + Good Learning: Collected Essays on Video Games, Learning, and Literacy*. 2nd ed., Lang, 2013.

———. *Literacy and Education*. Routledge, 2015.

———. *Situated Language and Learning: A Critique of Traditional Schooling*. Routledge, 2004.

———. *Social Linguistics and Literacies: Ideology in Discourses*. 5th ed., Routledge, 2015.

———. *Unified Discourse Analysis: Language, Reality, Virtual Worlds, and Video Games*. Routledge, 2014.

Gee, James Paul, and Elisabeth R. Hayes. *Language and Learning in the Digital Age*. Routledge, 2011.

———. *Women as Gamers: The Sims and 21st Century Learning*. Palgrave Macmillan, 2010.

Glenberg, Arthur M. "What Is Memory For?" *Behavioral and Brain Sciences*, vol. 20, 1997, pp. 1–55.

Glenberg, Arthur M., and Vittorio Gallese. "Action-Based Language: A Theory of Language Acquisition, Comprehension, and Production." *Cortex*, vol. 48, 2012, pp. 905–22.

Glenberg, Arthur M., and David A. Robertson. "Indexical Understanding of Instructions." *Discourse Processes*, vol. 28, 1999, pp. 1–26.

Hattie, John, and Gregory C. R. Yates. *Visible Learning and the Science of How We Learn*. Routledge, 2013.

Hitt, Jack. *Bunch of Amateurs: Inside America's Hidden World of Inventors, Tinkerers, and Job Creators*. Broadway Books, 2013.

Levinson, Stephen C. *Presumptive Meanings: The Theory of Generalized Conversational Implicature*. MIT P, 2000.

Pylyshyn, Zenon W. *Computation and Cognition*. MIT P, 1984.

Reeves, Byron, and Clifford Nass. *The Media Equation: How People Treat Computers, Television, and New Media Like Real People and Places*. Cambridge UP, 1999.

Stanovich, Keith E. *Progress in Understanding Reading: Scientific Foundations and New Frontiers*. Guilford Press, 2000.

Vygotsky, Lev S. *Problems of General Psychology, Including the Volume* Thinking and Speech. Edited by Robert W. Rieber and Aaron S. Carton, *The Collected Works of L. S. Vygotsky*, vol. 1, New York: Plenum, 1987.

Wason, Peter. "Reasoning." *New Horizons in Psychology*, edited by Brian M. Foss, Penguin, 1966, pp. 135–51.

———. "Reasoning about a Rule." *Quarterly Journal of Experimental Psychology*, vol. 20, 1968, pp. 273–281.

8

Reconfiguring Learner Identities
in Writing Pedagogy

Suresh Canagarajah

Noting that accountability practices are defining literacy narrowly to suit standardized testing, while scholarship is developing more expansive understandings of writing and language, Canagarajah makes the case that we now treat language as a situated, hybrid, multimodal semiotic resource, transcending the grammatical structures of separately labeled languages. We need to question distinctions we make among students by dividing them according to whether they speak English as a native language or a second language, as a first language or a second language, as a citizen or an immigrant, all of which determine whether they are native or nonnative to English. There are theoretical and practical problems in making such distinctions. These concerns are especially pronounced in the context of globalization and multilingualism.

While assessment and accountability practices are defining literacy narrowly to suit standardized testing, research and scholarly practices are developing more expansive understandings of writing and language. We now treat language as a situated, hybrid, multimodal semiotic resource, transcending the grammatical structures of separately labeled languages. How would such a notion of language redefine assessment and accountability in writing? The time has come to question a distinction we make among students in structuring our writing programs. Students are divided according to whether they speak English as a native language (ENL) or a second language (ESL); whether as a first language (L1) or a second language (L2); whether they are citizens or immigrants—all of which determine whether they are native or nonnative to English. There are theoretical and practical problems in making such distinctions. These concerns are especially pronounced in the context of globalization and multilingualism.

To begin with, it is difficult to enumerate one's language repertoires based on proficiency or time of acquisition. I grew up in a postcolonial household in Sri Lanka where English and Tamil were spoken, thus acquiring both languages from my infancy. I also played with Muslim neighbors who spoke a version of Arabic and Buddhist neighbors who spoke Sinhala, picking up their languages as well. I could be considered native[1] of four languages. It is difficult for me to distinguish which language was learnt first in my repertoire. I was socialized into these languages from infancy, making it difficult to say which is L1, L2, L3, and L4. Even based on proficiency, I find it difficult to decide on one as more important than the others, as I have relatively more proficiency in certain languages for certain skills. I speak more fluently and often in Tamil, while writing more fluently and often in English.

However, I cannot be considered a native speaker of four languages, as nativity is defined in terms of a single language one acquires since infancy in a homogeneous environment. Bloomfield states, "The first language a human being learns to speak is his native language; he is a native speaker of this language" (43). Chomsky adds the condition that this should be the only language spoken in a homogeneous environment, thus excluding bilinguals: "We exclude, for example, a speech community of uniform speakers, each of whom speaks a mixture of Russian and French (say, an idealized version of the nineteenth-century Russian aristocracy)" (17). From this perspective, even Anglo-Americans who might claim to be speakers solely of English cannot be considered native to it. This is because there is no homogeneous environment with only one language. As Mary Louise Pratt has argued, we inhabit contact zones where languages and cultures always interact (see Canagarajah, "ESL Composition," for a pedagogical exploration of contact zones). Anglo-Americans are surrounded by input from diverse languages in social and digital spaces that influence their English. They might be developing at least receptive proficiency in other languages. Furthermore, they switch between different registers and discourses, sometimes mixing in other languages. In fact, English is already a "creole" language, accommodating words and grammars from different languages throughout its history (see Fennell). Therefore, nativity depends on a myth of homogeneity that prevents us from appreciating the diversity in the repertoire of native speakers of English. From all these perspectives, there is the realization that the native speaker construct is an ideology, labeled "native-speakerism" (Holliday and Aboshiha), constituting a monolingual orientation that treats languages as autonomous and territorialized, conferring power on those who consider themselves the owners of a language.

There are detrimental consequences for identity deriving from the labels listed above. The terms *nonnative*, *L2*, or *ESL* identify learners according to a single scale of reference—namely, their relative proficiency in English. However, this is a deficient identification when we consider the fact that multilingual learners of English are bringing with them proficiency in other languages. They are bringing many linguistic and educational resources from their repertoires that these labels don't acknowledge. Additionally, native-speakerhood is treated as the norm against which a learner's proficiency is measured. However, multilingual learners may say that native speaker identity is not what they are aspiring to achieve. As for me, I aspire to be a good multilingual speaker of English, who integrates English into my other repertoires and appropriates it for my own voice and interests. I embrace the differences in my English as part of my identity. Therefore, native-speakerhood is an inappropriate and unfair frame of reference to judge the competence and use of multilinguals. Furthermore, these identity labels are rigid, as one can never cross the line from native to nonnative. Native-speakerhood is a birthright, often associated with racialized attributes. However long I learn English and develop advanced grammatical competence, English will never be considered native to me given my race and geographical background. Other non-Caucasian subjects have written about being treated as nonnative based on their skin color and physical features, despite the fact that English was the only language they spoke (see Lee's experience as a Chinese Canadian, in Canagarajah and Lee).

A final problem with these labels relates to their incongruity with the nature of writing. If writing is a rhetorical and multimodal activity, we have to question the importance given to language in identifying and separating students for writing instruction. Language is only one among many semiotic resources that go into text construction and literate interaction. If we separate students according to their language identity and design pedagogies based on their nonnative status, we might unwittingly convey the message that writing proficiency is determined by one's language. That message goes counter to how writing and literacy are defined in more ecological, situated, and multimodal ways by leading scholars in the profession (see Bazerman's and Wysocki's chapters in this collection). In fact, analogously, it is such an expansive, holistic, and plural orientation to language that we need in writing instruction.

Emerging Orientations to Language

What is emerging as the *translingual orientation* focuses on the synergy among languages that generates new grammars and meanings. Most scholars adopting this orientation avoid using the more common term *multilingual*, as it has been traditionally conceived as multiple languages enjoying their separate identity and structure even in contact. Therefore, they refer to this orientation in pejorative terms as "parallel monolingualisms" (Heller) or "two solitudes" (Cummins). Furthermore, in the translingual orientation, diversity is the norm, with speakers bringing different semiotic resources to the same interaction. In order for communication to succeed despite this diversity, interlocutors adopt situated negotiation strategies. In this sense, the monolingual orientation is turned upside down, shifting the emphasis from sharedness to diversity, grammar to practices, and cognition to embodiment. It must, however, be clarified that this orientation does have a place for shared norms evolving from practice. Repeated situated engagement leads to a sedimentation (Hopper) of language resources into new contact languages, which can become registers and dialects over time. Certain evolving norms are bolstered by language ideologies and treated as "standards." The key difference is that this bottom-up perspective on the constructed nature of norms allows for renegotiation and reconstruction in social interactions, which differs from the top-down perspective on stable and inflexible norms in the monolingual orientation.

There are significant implications for language competence and acquisition from the translingual orientation. Consider first how this orientation differs from models we have adopted in traditional language and literacy learning. The dominant and popular model perceives both languages in a bilingual's repertoire as "interfering" with each other in learning, as their relationship is perceived as conflictual. This orientation to language acquisition has come to be known as "subtractive," as the proponents hold that the new language would suppress or override the other in order for learning to be successful. On this premise, parents and teachers have sometimes insisted on discontinuing the use of heritage languages in order for children to master English. However, over time, linguists perceived that the languages don't have to be conflictual and can instead exist side by side. In this "additive" model, it was theorized that one can build a second language competence in addition to the first. However, the two languages were still perceived as distinctive and enjoying their autonomy, as if the learner had compartmentalized competences for each language.

The relationship between languages can be more complex and dynamic. In the "recursive" model, the languages in one's repertoire are treated as enabling the learning of each other, reconfiguring the competence of each in complex ways. To illustrate from my experience, when I learnt English in Sri Lanka, local schools didn't have a distinctive curriculum for writing. Writing (whether in Tamil or English) was part of whole-language pedagogy, integrated with speaking and reading. The essays we wrote were largely descriptive, narrative, or expository, with teacher feedback (usually a terminal comment) focusing on their rhetorical effectiveness. It was when I came for graduate studies to the United States that I developed my proficiency in the genre of academic writing, with an understanding of this as having its own rules and structures. When I moved back to Sri Lanka for teaching, I was under pressure to write academic articles in Tamil in order to share my knowledge with local scholars. I found it useful to adopt the genre of English academic essays with an anticipatory introductory paragraph that spelt out the thesis and well-organized body paragraphs with their clearly defined topic sentences. Some appreciated the way this writing inserted a difference into the Tamil discourse, reconfiguring my (and others') repertoire of Tamil academic genres. Over time I encountered some resistance to this writing. Readers commented that they found my style too calculated and explicit, underestimating their ability to interpret the meanings and presenting my ethos as condescending. I saw some value in their preference for indirection and experimented with adopting it in my Tamil writing. I am now drawing from Tamil oral (conversational) traditions and adopting more indirectness, narrative, and personal voice in my English academic writing. What this example suggests is the recursive nature of my writing skills acquisition. As I move forward with one language, I am reconfiguring my other language—whose features I then use to reconfigure my second language. Furthermore, we also see that language learning is not a one-shot deal. Certain skills, registers, and genres in a language are mastered at different times, and they can influence existing repertoires.

However, even the recursive model doesn't go far enough in accommodating the complexity of the translingual orientation. Consider the translingual model. It is different from the previous three models in at least four ways:

1. The languages are not treated as separate. One language fuses into the other. The model captures the fact that it is difficult to define where one language ends and the other begins. Consider lexical items that get appropriated and transformed as they travel from one community to another.

2. Whereas the other models present acquisition as linear, the translingual model presents acquisition as multidirectional, with influences from languages working on each other in multiple ways.

3. Whereas the other models present different languages (or skills) as having their own competence, the translingual model presents competence as integrated, with all languages in one's repertoire making up a synthesized language competence. People don't have different cognitive compartments for different languages.

4. Whereas the other models present competence as progressive and even complete for each language, the circulatory nature of acquisition in the translingual model presents competence as never ending. It is difficult to decide at what point a language has been completely learnt. Besides, as we acquire new linguistic and semiotic resources, our competence is constantly reconfigured, requiring new learning. Can even native speakers say that they are fully proficient in English and not feel challenged by new genres and registers that they need to master afresh?

The translingual orientation to language acquisition is not new. Though the dominant paradigm, labeled "cognitive-lingual" (Ortega), does perceive language acquisition as occurring one at a time, with cognition as the locus of this knowledge, other scholars have pointed out its limitations. Already in 1989, neurolinguist Grosjean cautioned his colleagues with an article perceptively titled "Neurolinguists, Beware! The Bilingual Is Not Two Monolinguals in One Person." In the field of teaching English to speakers of other languages (TESOL), British linguist Cook has been arguing since 1991 against the separation of languages in the acquisition process of English-language learners. He coined the phrase "multicompetence" to challenge Chomsky's notion of competence for single languages. His term captures the idea that people multitask or parallel process with their languages, not keeping them disconnected when they are learning or using them. Bilingual studies scholar Ofelia García summarizes developments in her field in relation to language and literacy acquisition as follows:

> Rather than focusing on the language itself and how one or the other might relate to the way in which a monolingual standard is used and has been described, the concept of translanguaging makes obvious that there are no clear-cut boundaries between the languages of bilinguals. What we have is a languaging continuum that is accessed. (47)

Uptake in the Writing Profession

It is important to examine how writing scholars and practitioners are taking to the alternative orientation to language and literacy acquisition. Their questions and concerns can help us clarify the implications for pedagogy. Scholars in L1 composition have been most receptive to the translingual orientation. Horner and Trimbur initially questioned the focus on a single language/dialect in writing pedagogy, cautioning us that native speaker students may not be served well in the context of globalization, which requires multilingual literacies for social and economic success. A later opinion piece in *College English* calls for a reconsideration of all aspects of writing (that is, error correction, pedagogy, curriculum, teacher development, and language policy) from the translingual orientation (Horner et al.). In response to their theoretical introduction, practitioners often ask how they might implement the translingual orientation for native speaker students who claim proficiency in only one language.

The distinction between "difference-in-similarity" and "similarity-in-difference" (Canagarajah, *Translingual Practice* 9) helps answer this question. The latter notion refers to the fact that in writing products that overtly feature diversity, there are processes that are similar. I have elsewhere demonstrated how the mixing of different codes by students from different language backgrounds still adopts certain common "strategies" of reading and writing (Canagarajah, "Negotiating"). Difference-in-similarity, on the other hand, captures the notion that texts that look similar contain features of difference that we don't easily perceive (or simply ignore) because of dominant purist and monolingual ideologies. If we perceive English as a creole language, there is diversity within what we call "English" or even "Standard English." Such diversity might gain special resonance in certain contexts of writing. *Pariah* might be a borrowing from Tamil that is now perfectly embedded in Standard English and won't be flagged as code switching or ungrammatical—as in the following occurrence in a recent news headline: "Obama, the pariah president" (Milbank). However, the rhetorical significance of the borrowing and additional layers of meaning stand out prominently when we consider that blacks have theorized their status in relation to caste inequalities in South Asia (DuBois). Note, furthermore, that Standard English is not a monolithic construct. Different writers provide different renditions of Standard English in relation to their own voices. Standard English accommodates immense diversity by featuring different registers and discourses. Therefore, L1 students should become sensitive to such diversity within Standard English and the similar strategies they adopt to encode and decode difference, as multilingual students do. Scholars in L1 writing have published on classroom practices to further such translingual awareness (see Bizzell; Sohan; Lu; Lu and Horner).

Though the adoption of translingual orientations in L1 writing is recent, it has a longer history in bilingual studies. Since 1971 the *Common European Framework of Reference for Languages* has been formulating "plurilingual" competence as the target for all learners in continental Europe. In line with this thinking, students are encouraged to develop literacy in different languages for different functions. Students in one class might learn history in Spanish, geography in French, and philosophy in German. The idea is that languages can be complementary, rather than redundant. In the United States, scholars like Ofelia García and Kris Gutiérrez have been developing parallel literacy in both English and Spanish for young learners, demonstrating that one language can complement the other and sometimes be fused in the same hybrid text (as exemplified in the writings of Gloria Anzaldúa).

The field of TESOL has also begun to reconsider its pedagogies in relation to the translingual orientation. After a symposium at the 2008 TESOL convention, in which senior scholars like Joshua Fishman, Jim Cummins, and Robert Phillipson considered how the learning of English cannot be separated from the other languages students bring with them (Taylor), there was a *TESOL Quarterly* special topic issue titled "Plurilingualism in TESOL" (September 2013) that featured studies from around the world on reconfiguring pedagogical practices in line with a translingual orientation. Teachers of L2 writing/literacy are part of this group of professionals and have published their classroom experiences in TESOL publications (see Amicucci and Lassiter; Jain; Lee; Marshall and Moore; Sayer).

The most explicit resistance to the translingual orientation so far has come from the community of second-language writing (SLW) professionals. It is important to represent their voices in order to address their pedagogical concerns. Some SLW scholars feel that the translingual orientation might distract students from the basics of Standard English that they need for academic and social success. Ruecker gives voice to these concerns in a brief footnote in his recent article in *College Composition and Communication*:

> I refrain from advocating recently popularized translingual pedagogies because of my concern that this movement may do students a disservice in a few different ways, namely by ignoring or misrepresenting a rich history of second language writing knowledge (Matusda, "Lure") and by possibly delaying students' attempts to learn standardized language varieties. I recognize that it is important for teachers to validate students' multiple language resources in classrooms and for researchers to challenge the privileging of standardized varieties in areas like assessment; however, students like those in my study typically enter college classrooms with a clear purpose: to learn a privileged standardized variety of English. When students have busy lives outside the classroom and have much to learn to increase their academic fluency, it is important to be cautious when encouraging the use of a pedagogical strategy being uncritically pushed by many without the requisite expertise in the processes of language acquisition. (116)

Ruecker voices at least three major concerns that require clarification. First is the view that translingual pedagogy might distract or delay students in their need to learn "a privileged standardized variety of English." The translingual orientation doesn't ignore standardized varieties of English (as I discussed above). Such norms are a social fact and can be ignored only at one's peril. What translingual pedagogies favor is deconstructing the notion of Standard English to make students aware that it is an arbitrary social construct put together through a lot of ideological work. Asif Agha painstakingly details how a particular regional dialect in South East England was made to index privilege and standard through ideological indoctrination via style manuals, newspapers, and dictionaries over many centuries. Teaching Standard English while making students aware of its ideological construction is not inconsequential to pedagogy. It helps students understand that they can critically engage with this variety to represent their voices and renegotiate its norms. If Standard English is treated as a stable, preconstructed, overpowering norm, students can be intimidated. Worse still, they can adopt it mechanically, feeling that they don't have spaces for creativity.

We must also be aware that Standard English can be taught even more effectively while having students shuttle between different dialects or language repertoires. There is anecdotal evidence that some appreciate the norms better when they find opportunities to deviate from them. Also, as they move between different languages, learners develop a keen sensitivity to the norms operative in specific communicative contexts. Presenting only one grammatical structure repeatedly is often considered efficient in product-oriented and behaviorist pedagogies. However, such grammar teaching has been exposed for its limited outcomes (Hartwell). Alternatively, pedagogies that teach norms while allowing students to shuttle between codes and contexts adopt a more practice-oriented approach, which might be more effective in the long run. To this argument, Ruecker would raise his second criticism. He feels that L2 students don't have the time for such learning as they "have busy lives outside the classroom and have much to learn to increase their academic fluency." But this considers multilingual students as having the capacity to learn only one language at a time. The assumption that learning multiple dialects or codes simultaneously is difficult is a monolingual bias. Multilingual students already bring with them experiences of and resources for mastering multiple semiotic systems outside the classroom (in fact, without teacher help). Shuttling between codes to acquire diverse norms is something that they always do in their multilingual communities (see Khubchandani).

Ruecker raises a third concern, which is that Standard English is what students want to acquire. Though it is true that everyone is under pressure to master privileged dialects for economic, educational, and social success, we also have to teach students where this desire comes from (Motha and Lin). We have to discuss the dominance of monolingual ideologies that make these privileged dialects appear natural and given. Besides, giving in uncritically to this desire will result in students losing their heritage languages and the diverse repertoires they bring with them. In response, as I discussed above, teachers of bilingual students are developing pedagogies that make spaces for their community repertoires and Standard English at the same time.

Moving to theoretical concerns, some in SLW consider the translingual orientation a mere fad that will die quickly. In a recent article Matsuda argues that the term *translingualism* is not necessary as linguistics has already addressed this notion with the term *multilingualism* (480). My introduction above to the differences between the two terms suggests otherwise. Based on this assumption of overlap, Matsuda goes on to argue that such novel intellectual positions always present earlier theories as wrong or limited in order for scholars to start their own bandwagon. Matsuda is perhaps forming this opinion from the fact that composition scholars have has only recently encountered the translingual perspective in writing scholarship. However, this is a narrow view of the history of this orientation. As outlined above, it is the monolingual orientation that is recent (dated by scholars as originating in seventeenth-century European modernity). The translingual orientation, as a practice, is longer and ever present in history, even in the West, despite the ideological dominance of monolingualism. Consider Trimbur's study of the formation of the English Only movement in the United States. While the founding fathers adopted a policy of expediency that subtly favored English, a vibrant translingualism formed around "a polyglot and multiethnic multitude that emerged through the very energies of mercantilism aboard ships and in port cities, in the slave castles of West Africa and on the New World plantations, and in pan-Indian resistance movements" (27). Moreover, while translingualism may be a fad for some scholars, it is a fact of life for millions of people in Asia, Africa, and South America, as they have struggled for centuries to preserve such language and literacy practices. Often during colonization, drastic efforts were taken by Europeans to stamp out these literacies, which differed from the word-based and monolingual literacies they promoted (see Baca for Mexico; Khubchandani for India; Makoni for Africa). For such people, the translingual orientation is far from being a fad. They will continue to fight for preserving their translingual literate and communicative practices whether the academic community appreciates them or not (Canagarajah, "Reconstructing").

Matsuda critiques one of my classroom-based studies (Canagarajah, "Negotiating") to substantiate his position. Taking examples from one of my student's writing, Matsuda focuses on the mixing of Arabic and visuals in her English essay to point out that I am engaging in a form of "linguistic tourism" (482) designed to titillate readers with language diversity. Matsuda fails to note that the study was conducted precisely to move the discussion beyond product to strategies, this point framing the study, with the article structured according to the reading/writing strategies employed by students to negotiate their texts. In other words, Matsuda homes in on features of the written product to extricate them from the discussion of strategies they are supposed to illustrate. More importantly, while there are thirteen other students (including native speakers) who demonstrate subtle forms of voice (including difference-in-similarity), Matsuda uses only one Arabic student to make his argument, ignoring the less dramatic resources and strategies of all students. My description of the strategies emerges from my teaching practice and contributes to further refining the pedagogy, which SLW scholars seek.

Behind such criticism is a concern of SLW scholars about their professional identity in the context of translingual pedagogies. To that end, they recently published an open letter to clarify the distinction between L2 writing and translingual writing (Atkinson et al.). This could become an important document to push forward our thinking about these paradigms. However, the authors give more space to professional concerns, such as gaining more opportunities for their work in journals, at conferences, and on the job market. The letter would have been more constructive and informative for the profession if the authors had modeled it after the translingual position statement of Horner and coauthors by explaining its theoretical bases and pedagogical ramifications (Horner et al.). Though the authors list four publications in the appendix as clarifying this distinction, they actually discuss the purported limitations of a translingual orientation rather than comparing the two orientations.

It is important to read the open letter closely to glean whatever insights are possible into how these scholars perceive the difference between L2 and translingual writing. The authors begin, "There seems to be a tendency to conflate L2 writing and translingual writing, and view the latter as a replacement for or improved version of L2 writing. This is not consistent with our understanding of the field of L2 writing" (Atkinson et al. 384). However, their understanding of the field of L2 writing is not then explained. Perhaps the next paragraph provides the best clue for how they see the difference: "L2 writing is an international and transdisciplinary field of study that is concerned with any issues related to the phenomenon of writing in a language that is acquired later in life" (384). Since these notions are also not explained, we have to infer the differences. Perhaps the authors are thinking of L2 writing as learnt consciously after childhood language acquisition. It is true that some of us acquire writing in another language later in life, as in my case. However, we have to ask if all genres of writing are included in this definition. In my case I did practice certain creative and personal genres in English, before I learnt academic writing when I came to the United States for graduate studies. Many multilingual students come to American universities having used English on social media sites and in digital communication before they learn academic writing (see You, for an example). The authors have to specify if they are considering a specific genre of writing in this definition.

Even if we grant the late acquisition of academic genres in English, we have to acknowledge that it will be difficult to isolate this learning to one language alone. If we return to my own example, my learning of English academic writing in the United States reconfigured my genres of Tamil writing, and my Tamil oral rhetorical resources influenced my English academic writing. My acquisition of L2 academic writing was translingual. It is not clear how this learning can be isolated to English. Even when teachers focus only on English in their classrooms, learners are shuttling between languages in cognitive processes and writing practices. The authors also fail to specify what age "later in life" means. We know that scholars in bilingual studies have been developing writing competence in English among students from elementary school onward, treating Spanish as a resource for acquiring English (for examples, see García; Gutiérrez).

Later in the open letter, the authors go on to state, "Translingual writing has not widely taken up the task of helping L2 writers increase their proficiency in what might still be emerging L2s and develop and use their multiple language resources to serve their own purposes" (384). As I demonstrated earlier, there is an increasing body of work that does adopt a translingual orientation to help multilingual students develop their writing proficiency in English. However, the pertinent question here is why these studies and scholars are not treated as belonging to L2 scholarship. The issue is that these studies don't always appear in second-language studies journals or are framed within SLW studies (for the obvious reason that translingual studies favor broadening the discussion beyond the L1/L2 binary). However, it will serve the field better if the notion of L2 writing is expanded beyond those scholars typically featured in journals or conferences labeled SLW to include important scholars in TESOL, bilingual studies, and L1 composition (such as Ofelia García, Min-Zhan Lu, and Kris Gutiérrez) who are not only multilingual scholars themselves but are addressing the concerns of multilingual students and their writing in their own research and pedagogy. This inclusive approach will enrich L2 studies and disciplinary discourses.

The rest of the letter details professional concerns, which are again repeated in the conclusion to the letter. This concern for the future of the disciplinary community is certainly understandable. However, we can take a cue from scholars in L1, TESOL, and bilingual studies in the way they handle the new orientation to language and literacy in their work. They have taken on the new realizations and reconfigured their disciplinary constructs from within the field without fear of losing their disciplinary status. In fact, they consider the translingual orientation to enrich the work they do rather than hampering them. Of course, in the long run, we cannot guarantee what these challenging orientations to language and literacy may mean to all our fields. There are interesting debates on the future of humanities in general and English in particular, with scholars criticizing the narrow ways in which disciplinary boundaries have been drawn (Jay; Leitch). In the short run, however, L2 scholars can engage with new thinking to enrich their research and teaching.

Way Forward

How do we adopt the translingual orientation to make practical and meaningful changes in writing pedagogy? Though there are fascinating examples from the studies in L1 writing, bilingual literacies, and TESOL that I have cited above, I wish to take up a high-stakes policy proposal under national discussion. I consider how the Common Core State Standards Initiative (CCSSI; http://www.corestandards.org) envisions English literacy and how the standards are creatively adapted by practitioners in a specific state for their own purposes in alignment with their translingual orientation.

The CCSSI doesn't make a distinction between native and nonnative students in formulating its standards for secondary school students in preparing them for college. It is admirably egalitarian in holding both groups to the same standard and treating all students as having the competence to attain it. One doesn't see terms such as *native* and *nonnative* in its description of students or standards. However, the CCSSI does include an addendum titled "Application of Common Core State Standards for English Language Learners" to clarify its position toward nonnative students (which it labels English-language learners, or ELL). It states,

> The Common Core State Standards for English language arts (ELA) articulate rigorous grade-level expectations in the areas of speaking,

listening, reading, and writing to prepare all students to be college and career ready, including English language learners. Second-language learners also will benefit from instruction about how to negotiate situations outside of those settings so they are able to participate on equal footing with native speakers in all aspects of social, economic, and civic endeavors.

One should applaud the CCSSI's motivation to expect the same high standards and communicative competence for native and nonnative students whether in academic or social contexts.

The CCSSI is also motivated by enlightened orientations to the relationship between language and content. It says,

> The development of native like proficiency in English takes many years and will not be achieved by all ELLs especially if they start schooling in the US in the later grades. Teachers should recognize that it is possible to achieve the standards for reading and litera-ture, writing & research, language development and speaking & listening without manifesting native-like control of conventions and vocabulary.

In saying this, the CCSSI decouples surface-level issues such as accent and vocabulary from one's communicative practices in English. Often such super-ficial markers become signs of cognitive deficiency for teachers, who assign such students to special instruction. However, one's abilities in literacy and content-level instruction involve more complex communicative, multimodal, and thinking resources beyond grammatical control.

In line with this thinking, the CCSSI acknowledges the special resources that ELLs bring with them for academic and communicative success in En-glish-language environments:

> ELLs bring with them many resources that enhance their education and can serve as resources for schools and society. Many ELLs have first language and literacy knowledge and skills that boost their ac-quisition of language and literacy in a second language; additionally, they bring an array of talents and cultural practices and perspectives that enrich our schools and society.

In this manner, it adopts an additive orientation to bilingualism, considering the cultures and languages students bring with them as helping rather than hindering the English literacy development of these students.

The statement, therefore, goes on to make special provisions for ELLs to achieve the same standards as native students. It says, "However, these students may require additional time, appropriate instructional support, and aligned assessments as they acquire both English language proficiency and content area knowledge." The statement goes on to provide practical suggestions for how teachers can be made skillful in understanding the special needs of ELLs and serving them better. Though these provisions should be appreciated, some might also consider them a bit condescending. The provisions assume that these students don't have the capacity to master multiple languages and diverse knowledge constructs simultaneously. More importantly, it appears that the CCSSI is more interested in using the resources of ELLs to transition them to its own (English-based) standards rather than engaging with their resources equally and developing them at the same time. In other words, the L1 resources subserve the goals of the CCSSI (for a critique on this point, see Flores and Schissel). To refer to the models already described, the CCSSI is additive but not translingual.

While the CCSSI holds all students to the same expectations, it ignores that its standards will be realized by multilingual students differently, in relation to the knowledge and communicative resources they bring with them. In other words, it adopts a one-size-fits-all approach to its expected standards. There is no realization that different students will achieve those standards in different ways. Consider, for example, one of the general principles that guide the formulation of the CCSSI, as articulated in its overview:

> The Common Core emphasizes using evidence from texts to present careful analyses, well-defended claims, and clear information. Rather than asking students questions they can answer solely from their prior knowledge and experience, the standards call for students to answer questions that depend on their having read the texts with care. ("Key Shifts")

Even if students read the texts with care, presumably limiting themselves to a close reading within the narrow bounds of the text, their interpretation will be influenced (enriched?) by the prior knowledge and experience they bring with them. It is difficult to cut off that knowledge from any reading. Such influences will motivate multilingual students to come up with slightly different interpretations. Even "evidence" will be different in relation to what counts as effective in different communities. As sociolinguist Thomas Kochman once pointed out, what is ad hominem argument in the classical rhetorical tradition is valid evidence for African American and other oral communities who hold that the ethos, sincerity, and credibility of the presenters should be considered in assessing their position. If presenters make an argument that deviates from this practice (if they don't walk their talk, in other words), their argument loses credibility. Furthermore, the rhetorical features expected in presenting evidence can also be problematic. Careful analysis, good defense, and clarity can differ from community to community. As I mentioned already, my Tamil students in Sri Lanka criticized me for being too controlled and explicit in my writing. They considered indirection as "clear" for them, apart from providing me an equal footing with readers who felt their intelligence was being respected. From this point of view, then, the CCSSI treats its standards in a monolithic way.

However, the good news is that the CCSSI repeatedly reminds us that it is only formulating the standards while leaving the means to achieve them to local teachers and schools. This gives some leeway to teachers for creativity and agency. We have to remember that any policy document is a discursive construct that is open to interpretation. Language-policy scholars have studied how teachers interpret policy documents to their advantage. While the No Child Left Behind policy motivated some teachers to teach to the test, others considered creative and critical thinking as enabling their students to meet the expected standards (see Hornberger and Johnson). Similarly, the formulation of the CCSSI is being redefined by certain states in relation to their pedagogical priorities. I present, as an exemplary case, how New York State has redefined its pedagogical constructs as it meets the standards of the CCSSI (see resources available on the site EngageNY: https://www.engageny. org/ccss-library).

The document titled "NYS Bilingual Common Core Initiative: Theoretical Foundations" (NYSBCCI) adopts the translingual orientation as its theoretical premise in the opening two paragraphs. It adopts the term *dynamic bilingualism* to distinguish it from traditional forms of bilingualism that treat both languages in an additive relationship of two monolingualisms. In accordance with this orientation, it strikingly adopts new constructs to talk about students and proficiency levels. Having previously adopted levels such as *beginning, intermediate, advanced*, and *proficient*, it now uses five levels of proficiency: *entering, emerging, transitioning, expanding*, and *commanding*. Is there a difference? The previous stages largely evoke cognitive and evaluative categories, while the latter adopt temporal and, thus, pragmatic categories to classify students according to their exposure to and interaction with the language. Furthermore, the previous terminology posits the state of being "proficient" as an end point of learning. As I discussed earlier, it is difficult to define a threshold level for perfect proficiency in any language. The present participle, the grammatical form of the new labels, is a reminder that proficiency is ongoing.

Similarly, in place of constructs such as nonnative, ESL, L2, and ELL, which adopt a single scale and enumerative orientation to classify the status of one's English, NYSBCCI adopts the term *new language* in its core pedagogical focus: "New Language Arts Progressions." It explains, "Using *new* as opposed to *second* language [. . .] acknowledges the many students in New York State who have competency in more than two languages" (2). The second core pedagogical focus, aimed at developing parallel proficiency in what was traditionally considered the heritage or native language of the student, is titled "Home Language Arts Progressions." NYSBCCI explains the rationale for the change as follows:

> This change is aligned with recent developments in language education that find the term "native" speaker to be a concept no longer applicable to our increasingly globalized world (Graddol, 2006). In addition, the shift in terminology allows for a message that home and school can and must be integrated in ways that allow students to see them as complementary spheres as opposed to separate spheres in their lives. (3)

It is clear that the policy makers display the creativity and boldness to construct meaningful new labels that take into account the limitations of traditional constructs from a monolingual orientation. Beyond just changing labels, they also envision a bold pedagogy that connects the resources of home and school as integrated in learning English.

Their pedagogical vision is explained right after their theoretical adoption of the translingual orientation: "In line with this research, the NYSBBCI views bilingualism both as a point of departure for language instruction and as goal for all language learners" (1). This is a radical approach to English-language pedagogies. While the CCSSI treats bilingualism as a point of departure for transitioning to English, NYSBCCI goes further to treat it also as a goal for learning. In other words, it posits that the best way to learn English is by simultaneously developing proficiency in the home language of the students. This policy perceives languages as complementary rather than interfering, in which developing a higher-order and hybrid translingual proficiency is possible without failing to meet the standards of the CCSSI. This is a paradox but is well motivated by the translingual orientation, which acknowledges the multidirectional and ever-continuing influences of language resources in a learner's communicative capacity.

It is for this reason that NYSBCCI considers its two pedagogical foci of New Language Arts Progressions and Home Language Arts Progressions as interconnected. It explains,

> These tools are designed primarily to meet the needs of English Language Learners; however, to support a broader goal of bilingualism for all students, these resources can also be used as a guide for planning instruction for students who are learning a foreign language or who are developing their home languages. [. . .] The New Language Arts Progressions and Home Language Arts Progressions are not separate standards, but rather provide a roadmap for teachers to ensure that students who are learning a new language and/or developing their home language meet the Common Core standards. (1–2)

In integrating the different languages learnt, NYSBCCI sees that similar tools, standards, and progression apply to students when learning English, home languages (such as Spanish or Chinese), or foreign languages (such as Spanish for a Chinese student). It might be inferred that the same can be applied to native students who might be learning their forgotten heritage language (Italian) or a foreign language (Chinese). Thus this proposal takes pedagogy beyond the native/nonnative divide.

NYSBCCI goes on to deconstruct certain other myths and biases in language teaching. Should students first master the fundamentals of English grammar before they engage with subject-specific instruction? The policy clarifies: "When provided appropriate scaffolding, language learners can start developing language for academic purposes at the same time that they are developing basic communication skills in their new language (Walqui & Heritage, 2012)" (2). Here NYSBCCI is treating cognitive and knowledge processes as transcending grammatical proficiency. In fact, the policy makers see both content and grammar as facilitating each other in learning. This approach adopts a practice-based orientation in opposition to the traditional product-oriented view of teaching grammar separately from (and before) content and communicative practice. Similarly, should students learn to use English in basic conversational interactions before they can start writing? Does orality precede literacy? NYSBCCI states,

> A curriculum for these students must include all four components of language (listening, speaking, reading, and writing). [. . .] Organizing language development as productive and receptive ensures the integration of the four components of language and emphasizes that

> students who are new to a language do not need to first develop oral
> language before being exposed to written language. (3)

Here again it promotes an integrated curriculum that doesn't treat writing as a higher skill separated from others but as drawing from all modes of communication. SLW scholars who think there are no pedagogies to practice the translingual orientation should visit NYSBCCI's website to see the suggested activities and texts for grades K–12 (https://www.engageny.org/resource/new-york-state-bilingual-common-core-initiative).

To understand NYSBCCI's bold pedagogy, we must take into account its broader vision. Its concluding statement notes, "When used together, New Language Arts Progressions and the Home Language Arts Progressions provide a roadmap to develop bilingual Common Core skills for all students—skills that are necessary for our increasingly global society" (4). NYSBCCI is designed to provide students the resources needed for global citizenship. There is an implicit criticism that the CCSSI is constructed more in terms of national priorities and nation-state ideologies. It is NYSBCCI's transnational vision that explains how to develop translingual resources in the place of separate monolingualisms. Hence there is also the need to go beyond monolithic standards to acknowledge that these standards can be realized differently in different communities and contexts.

One must give credit to the policy makers and practitioners of this secondary school curriculum for boldness, imagination, and creativity in coming up with new constructs and pedagogies to address the emerging understanding of language in a global context. While colleges and universities are still struggling to go beyond the native/nonnative divide, secondary school teachers have taken bold steps to reimagine a new pedagogy for English literacy. Though NYSBCCI has its own limitations, with some scholars making the ideological criticism that it is still exploiting home-language resources to meet the monolithic standards of the CCSSI (see Flores and Schissel), I adopt the pragmatic view that scholars have to articulate how the translingual orientation can still meet context- or community-specific needs, norms, and agendas. NYSBCCI is an attempt to reconfigure the standards of the CCSSI from within and to show the relevance of a translingual orientation to meeting (in fact, exceeding) currently dominant standards.

Notes

An expanded and different version of this article appears in *Applied Linguistics Review*, December 2015.

1. Though I consider terms like *native* and *nonnative* to be myths, I continue to use them in this article (without quotation marks) as the distinction needs to be examined

closely before we use new terms (which I introduce at the end of the article). The same applies to "Standard English."

Works Cited

Agha, Asif. "The Social Life of Cultural Value." *Language and Communication*, vol. 23, 2003, pp. 231–73.

Amicucci, Ann, and Tracy Lassiter. "Multimodal Concept Drawings: Engaging EAL Learners in Brainstorming about Course Terms." *TESOL Journal*, vol. 5, no. 3, 2014, pp. 523–31.

"Application of Common Core State Standards for English Language Learners." Common Core State Standards Initiative, www.corestandards.org/assets /application-for-english-learners.pdf. Accessed 15 Apr. 2018.

Atkinson, Dwight, et al. "Clarifying the Relationship between L2 Writing and Translingual Writing: An Open Letter to Writing Studies Editors and Organization Leaders." *College English*, vol. 77, no. 4, Mar. 2015, pp. 383–86.

Baca, Damian. "Rethinking Composition, Five Hundred Years Later." *Journal of Advanced Composition*, vol. 29, 2009, 229–42.

Bizzell, Patricia. "Toward 'Transcultural Literacy' at a Liberal Arts College." *Reworking English in Rhetoric and Composition: Global Interrogations, Local Interventions*, edited by Bruce Horner and Karen Kopelson, Southern Illinois UP, 2014, pp. 131–49.

Bloomfield, Leonard. *Language*. Holt, Rinehart, and Winston, 1933.

Canagarajah, Suresh. "ESL Composition as a Literate Art of the Contact Zone." *First-Year Composition: From Theory to Practice*, edited by Deborah Coxwell-Teague and Ronald Lunsford, Parlor Press, 2014, pp. 27–48.

———. "Negotiating Translingual Literacy: An Enactment." *Research in the Teaching of English*, vol. 48, no. 1, 2013, pp. 40–67.

———. "Reconstructing Heritage Language: Resolving Dilemmas in Language Maintenance for Sri Lankan Tamil Migrants." *International Journal of the Sociology of Language*, no. 222, 2013, pp. 131–55.

———. *Translingual Practice: Global Englishes and Cosmopolitan Relations*. Routledge, 2013.

Canagarajah, Suresh, and Ena Lee. "Negotiating Alternative Discourses in Academic Writing: Risks with Hybridity." *Risk in Academic Writing: Postgraduate Students, their Teachers and the Making of Knowledge*, edited by Lucia Thesen and Linda Cooper, Multilingual Matters, 2014, pp. 59–99.

Chomsky, Noam. *Knowledge of Language: Its Nature, Origin, and Use*. Praeger, 1986.

Common European Framework of Reference for Languages. Council of Europe, www. coe.int/en/web/common-european-framework-reference-languages. Accessed 15 Apr. 2018.

Cook, Vivian. "The Poverty-of-the-Stimulus Argument and Multi-competence." *Second Language Research*, vol. 7, 1991, pp. 103–17.

Cummins, Jim. "Teaching for Transfer: Challenging the Two Solitudes Assumption in Bilingual Education." *Encyclopedia of Language and Education*, edited by Cummins and Nancy Hornberger, vol. 5, Springer, 2008, pp. 65–75.

DuBois, W. E. B. "Evolution of the Race Problem." *Proceedings of the National Negro Conference.* National Negro Conference, 1909, pp. 142–58.

Fennell, Barbara. *A History of English: A Sociolinguistic Approach.* Blackwell, 2001.

Flores, Nelson, and Jamie L. Schissel. "Dynamic Bilingualism as the Norm: Envisioning a Heteroglossic Approach to Standards-Based Reform." *TESOL Quarterly*, vol. 48, no. 3, 2014, pp. 454–79.

García, Ofelia. *Bilingual Education in the 21st Century: A Global Perspective.* Wiley-Blackwell, 2009.

Grosjean, François. "Neurolinguists, Beware! The Bilingual Is Not Two Monolinguals in One Person." *Brain and Language*, vol. 36, 1989, pp. 3–15.

Gutiérrez, Kris. "Developing a Sociocritical Literacy in the Third Space." *Reading Research Quarterly*, vol. 43, no. 2, 2008, pp. 148–64.

Hartwell, Patrick. "Grammar, Grammars, and the Teaching of Grammar." *College English*, vol. 47, no. 2, 1985, pp. 105–27.

Heller, Monica. *Linguistic Minorities and Modernity: A Sociolinguistic Ethnography.* Longman, 1999.

Holliday, Adrian, and Pamela Aboshiha. "The Denial of Ideology in Perceptions of 'Nonnative Speaker' Teachers." *TESOL Quarterly*, vol. 43, no. 4, 2009, pp. 669–89.

Hopper, Paul. "Emergent Grammar." *Berkeley Linguistics Society*, vol. 13, 1987, pp. 139–57.

Hornberger, Nancy, and David Johnson. "Slicing the Onion Ethnographically: Layers and Spaces in Multilingual Language Education Policy and Practice." *TESOL Quarterly*, vol. 41, no. 3, 2007, pp. 509–32.

Horner, Bruce, et al. "Opinion: Language Difference in Writing: Toward a Translingual Approach." *College English*, vol. 73, no. 3, 2010, pp. 303–21.

Horner, Bruce, and John Trimbur. "English Only and U.S. College Composition." *College Composition and Communication*, vol. 53, 2002, pp. 594–630.

Jain, Rashi. "Global Englishes, Translinguistic Identities, and Translingual Practices in a Community College ESL Classroom: A Practitioner Researcher Reports." *TESOL Journal*, vol. 5, no. 3, 2014, pp. 490–522.

Jay, Paul. "Beyond Discipline? Globalization and the Future of English." *PMLA*, vol. 116, no. 1, 2001, pp. 32–47.

"Key Shifts in English Language Arts." Common Core State Standards Initiative, www.corestandards.org/other-resources/key-shifts-in-english-language-arts/. Accessed 15 Apr. 2018.

Khubchandani, Lachman. *Revisualizing Boundaries: A Plurilingual Ethos.* Sage, 1997.

Kochman, Thomas. *Black and White Styles in Conflict.* U of Chicago P, 1981.

Lee, Melissa. "Shifting to the World Englishes Paradigm by Way of the Translingual Approach: Code-Meshing as a Necessary Means of Transforming Composition Pedagogy." *TESOL Journal*, vol. 5, no. 2, 2014, pp. 312–29.

Leitch, Vincent B. "Postmodern Interdisciplinarity." *Profession*, 2000, pp. 124–31.

Lu, Min-Zhan. "Metaphors Matter: Transcultural Literacy." *Journal of Advanced Composition*, vol. 29, 2009, pp. 285–94.

Lu, Min-Zhan, and Bruce Horner. "Translingual Literacy, Language Difference, and Matters of Agency." *College English*, vol. 75, no. 6, 2013, pp. 582–607.

Makoni, Sinfree. "From Misinvention to Disinvention: An Approach to Multilingualism." *Black Linguistics: Language, Society, and Politics in Africa and the Americas*, edited by Geneva Smitherman et al., Routledge, 2002, pp. 132–53.

Marshall, Steve, and Danièle Moore. "2B or Not 2B Plurilingual? Navigating Languages, Literacies, and Plurilingual Competence in Postsecondary Education in Canada." *TESOL Quarterly*, vol. 47, no. 3, 2013, pp. 472–99.

Matsuda, Paul Kei. "The Lure of Translingual Writing." *PMLA*, vol. 129, no. 3, 2014, pp. 478–83.

Milbank, Dana. "Obama, the Pariah President." *The Washington Post*, 20 Oct. 2014, www.washingtonpost.com/opinions.

Motha, Suhanthie, and Angel Lin. "'Non-coercive Rearrangements': Theorizing Desire in TESOL." *TESOL Quarterly*, vol. 48, no. 2, 2014, pp. 331–59.

"NYS Bilingual Common Core Initiative: Theoretical Foundations." *Engage NY*, New York State Education Department, www.engageny.org/resource/new-york -state-bilingual-common-core-initiative/file/135506. Accessed 15 Apr. 2018.

Ortega, Lourdes. "Ways Forward for a Bi/Multilingual Turn in SLA. *The Multilingual Turn: Implications for SLA, TESOL, and Bilingual Education*, edited by Stephen May, Routledge, 2014, pp. 32–52.

Ruecker, Todd. "Here They Do This, There They Do That." *College Composition and Communication*, vol. 66, no. 1, 2014, pp. 91–119.

Sayer, Peter. "Translanguaging, TexMex, and Bilingual Pedagogy: Emergent Bilinguals Learning through the Vernacular." *TESOL Quarterly*, vol. 47, no. 1, 2013, pp. 63–88.

Sohan, Vanessa Kramer. "Working English(es) as Rhetoric(s) of Disruption." *Journal of Advanced Composition*, vol. 29, nos. 1–2, 2009, pp. 270–75.

Taylor, Shelley. "Paving the Way to a More Multilingual TESOL." *TESOL Quarterly*, vol. 43, no. 2, 2009, pp. 309–13.

Trimbur, John. "Linguistic Memory and the Uneasy Settlement of U.S. English." *Cross-Language Relations in Composition*, edited by Bruce Horner et al., Southern Illinois UP, 2010, pp. 21–41.

You, Xiaoye. "Chinese White-Collar Workers and Multilingual Creativity in the Diaspora." *World Englishes*, vol. 30, no. 3, 2011, pp. 409–27.

9

Creating a United Front: A Writing Program Administrator's Institutional Investment in Language Rights for Composition Students

Staci Perryman-Clark

Three campus-wide retention initiatives at Western Michigan University were developed to support speakers of non-Western languages or English language varieties. The author addresses theoretical and practical problems associated with distinctions based on whether or not students are English native speakers, English as a second language speakers, first-language speakers, or second-language speakers, by focusing on students in these linguistic categories to appeal to the institutional administrators who make granting decisions. The author concludes by considering the implications of framing such initiatives in the context of Students' Right to Their Own Language, a position statement of the Conference on College Composition and Communication, and suggests how composition teachers and writing program administrators might strategically and tactically garner similar support for such efforts on their own campuses.

In the introduction to the edited collection *Students' Right to Their Own Language: A Critical Sourcebook*, Staci Perryman-Clark, David E. Kirkland, and Austin Jackson examine the vast breadth of scholarly conversation concerning "one of the most controversial position statements" that the Conference on College Composition and Communication, or CCCC, has passed in its history: Students' Right to Their Own Language, or SRTOL (1).[1] While there have been significant developments, empirical research, and pedagogical strategies used to support teachers and students in applications of SRTOL (Smitherman; Richardson and Gilyard; Richardson; Perryman-Clark), the question remains to what extent SRTOL specifically supports all groups affected by language differences and differences in language variation. Perryman-Clark and coauthors specifically ask,

> How does SRTOL support ESL, EL, ESD, L1, L2, and bilingual
> students in the classroom? How does it include non-English language
> varieties, such as the Spanishes, Mandarins, Arabics, First Nation
> Languages, and so on, that have increasingly become part of the
> linguistic stew of American Languages? (6)

In response to this question I will discuss the initiatives I have developed with regard to linguistic diversity and SRTOL as a writing program administrator (WPA) at Western Michigan University (WMU). These initiatives were designed as retention initiatives to respond to high failure and withdrawal rates for speakers and writers of Arabics, Spanishes, and African American Language.

In this chapter I will share three campus-wide retention-related initiatives that seek to support speakers of non-Western languages or English language varieties. Like Suresh Canagarajah writes in his chapter for this collection, I agree that "there are theoretical and practical problems" associated with making distinctions based on whether or not students are English native speakers, English as a second language (ESL) speakers, first-language (L1) speakers, or second-language (L2) speakers; however, I have chosen to identify students with these categories given my institutional context and the audience of administrators on which I relied to receive financial and institutional support. Like James Paul Gee writes, also in this collection, I too understand that "language is crucial to any discussion about writing" and, therefore, have employed a focus on language as a rhetorical tactic for proposing initiatives about writing to the upper administrators who are frequently charged with making decisions about what initiatives to fund that could lead to student success. I conclude by sharing the implications of framing these initiatives in relation to the CCCC's SRTOL and by suggesting how composition teachers and WPAs might strategically and tactically garner support for such efforts on their own campuses.

When discussing my three campus-wide retention-related initiatives, I focus on retention because, on our campus, there is a clear relationship between success or failure and linguistic practices for students who take our first-year writing course, ENGL 1050: Thought and Writing. The initiatives I describe in this chapter were each framed with regard to student success, diversity, and retention because this framing has garnered the most funding and support from our upper administration, a framing that I often characterize as a tactic due to my subject position as an untenured (at the time) African American female WPA. These initiatives are ongoing with regard to their research and funding.

The first initiative, our ENGL 1050 rescue section initiative, is designed to give students who find that they are failing by midterm a second chance at passing ENGL 1050. The second initiative, a diversity, equity, and inclusion grant, was an internally funded project in collaboration with WMU's writing center to confront cultural biases associated with plagiarism across different writing courses at WMU. This initiative funded speakers and professional development support for teachers of L2 and ESL students. At WMU we've found that these students are reported for plagiarism practices disproportionately more often than mainstream U.S. domestic students even though mainstream U.S. domestic students commit more acts of plagiarism. The consequences of being reported for plagiarism have led to dismissal or withdrawal from WMU for some of the students reported. The third and final initiative for ENGL 1050: Thought and Writing (also in collaboration with our campus writing center) was a project cosponsored by both WMU and the Kellogg Foundation as part of a racial healing grant. For this grant we proposed to assess L2 students in ENGL 1050 to determine the needs, resources, and support that such learners might need. The commonality that exists with each of these initiatives is that (1) WMU instructors have traditionally operated from a theoretical assumption that student failure results from deficit, and (2) this assumption of deficit behind student failure has often influenced negative assumptions about students who write in their home languages, particularly when these languages are other than English. In short, the discourse of failure is taken up with the discourse of having an alternative language variety. As a WPA I've worked on crafting initiatives to remedy these troubling assumptions and biases.

ENGL 1050 Rescue Section Initiative and the Rhetoric of Deficit

This story frames the beginning of our ENGL 1050 rescue section initiative. In December 2013 I found myself in the middle of a dispute between an African American male ENGL 1050 student and a white female first-time teaching assistant (TA). The TA had previously reported the student for plagiarism on an assignment because, according to her, his *own language* differed drastically from many of the other texts that may have been patch written into the draft he submitted. The TA also claimed that the student had been intimidating and disrespectful toward her during several meetings on this subject and other topics outside class. As a result I agreed to meet one-on-one with both the student and the instructor. While I observed disrespect and resistance in his attitude (his defensive disposition was understandable since he had just been accused of and reported to me for plagiarism), I had seen no evidence that he had been threatening or intimidating. The TA, nonetheless, requested that he be removed from her class because of her own fears. Both my department chair and I agreed to remove the student so that my chair (with my support) could work individually with the student, given that we were less confident that the instructor would grade him fairly. The student's work improved considerably over the course of my chair's five weeks of work with him, and all was well until the TA requested to assign the student's final grade, even though she had requested that the student be removed from her class. She determined that because the student was failing prior to being removed from her class that he should receive a D at best. The problem, however, was that the student needed a C or higher in the course to declare a major (and, therefore, to be *retained* by the university) or else risked dismissal from the university. Based on our assessment of the student's portfolio, we believed he should earn a C, but both the TA and her graduate advisor challenged our beliefs. The challenge went to the dean of our college, with attempts to contact the provost with the request that both my department chair and I be removed from our positions!

Fortunately, the administration sided with the department chair and me; our assessment of the student's work, in addition to our having taught college composition for many years compared with the TA's one semester of teaching, made the administration's decision an easy one. As a result of this conflict, my department chair wondered how many additional cases existed of instructors having negative assumptions and biases toward their students' language and cultural and racial identities. We decided, then, that it would be better to create a formal opportunity to handle such cases when there were conflicts between instructors and students. We proposed a fall 2014 ENGL 1050 rescue section initiative, funded and with the support of two full-time faculty members with doctoral degrees who agree to teach two rescue sections that open for registration after midterm. Students formally apply with a written statement of how they plan to improve and perform better (see appendix A). They also are required to have a written endorsement from their current composition teacher concerning their potential to succeed, even though their average is currently a failing one at midterm. Once students are admitted, they complete a portfolio of revised work from their previous course along with an annotated bibliography, a research project, and a reflective piece, so that they are still completing the same amount of work as those enrolled in traditional sections. They also don't have to pay late registration fees or tuition for switching to rescue sections.

The trial run during the 2014–15 academic year had the following results:

- Thirty students applied and were admitted to our ENGL 1050 rescue sections.
- Of those thirty admitted, twenty-four students passed who would have otherwise failed ENGL 1050.

- Two-thirds of the students enrolled in these sections were ethnic minority or international students.
- One-fifth were classified as L2 students.
- Thirteen students are currently in academic good standing as of spring 2015.
- Seven students are on academic probation as of spring 2015.
- Eleven students are registered for the fall 2015 semester and therefore have been retained (LaHaie and Redding).

These findings have had noteworthy impact on the First-Year Writing Program at WMU. The program's number of students passing ENGL 1050 has increased 22 percent. The increase in passing has come mostly from at-risk populations of students, including ethnic minorities, international students, and working-class students. It is also worth pointing out that this initiative achieved a 66 percent retention rate, which is higher than the overall university average (LaHaie and Redding). Impressed by our retention efforts, we received support and funding to continue the program for at least one additional academic year. We do hope to increase the number of students we retain next year.

Based on this experience, we've learned to be more mindful of the problems associated with operating from a deficit assumption of failure when advocating retention initiatives. In *Retention and Resistance: Writing Instruction and Students Who Leave*, Pegeen Reichert Powell asserts, "Part of the failure of the discourse of retention is that it reinforces the traditional narrative of higher education by treating students who leave as individual problems to be solved, situations that require intervention" (100). Calling our initiative one of rescue, then, also treats students as problems or as those in need of rescue. As a result we are currently revising this opportunity by framing it as a writing-intensive section, and not a rescue one. The significance of this initiative, though, is that it reemphasizes the framing of many students who use other languages besides English as those who need to be remediated and rescued. However, when competing for internal funding, it is often retention, success, and failure that seem to be the only discourses that upper administrative officials understand. It is for this reason that we framed retention and failure as a tactic, in the de Certeauian sense, to garner funding and support.[2]

Scholarship in composition and literacy also points to the problems associated with framing language as both deficit and failure. As Canagarajah rightly points out in this collection, labeling students "academically deficient" whose languages or language varieties differ from that of native English speakers has severe consequences for those students. Canagarajah notes that "there are detrimental consequences for identity deriving from the labels . . . *nonnative, L2,* or *ESL* [that] identify learners according to a single scale of reference—namely, their relative proficiency in English." Thus, the degree to which a multilingual learner is considered proficient is often constructed by the instructor who makes biased assumptions about the student's performance and success based on how (s)he labels and classifies the student as a native English speaker or an ESL, L2, or nonstandard-language-variety speaker.

On our campus we noticed that, initially, instructors primarily referred U.S. domestic white students to the rescue sections during fall semester 2014; however, more instructors referred African American students, most of whom were speakers of African American Language, and ESL and L2 students during spring semester 2015. Students from these backgrounds, then, accounted for two-thirds of the thirty students who enrolled in rescue sections during the 2014–15 academic year. We also noticed that during spring semester 2015, the total number of applicants was twice what it had been in fall 2014 (Redding and LaHaie). Furthermore, during spring 2015, when we included a more specific referral questionnaire about English-language problems that asked instructors what current issues students were having in the course (see appendix B), most instructors interpreted this question as justification for referring failing U.S. racial minority and international students (whom most instructors labeled as ESL) to our rescue sections. They connected ESL and ethnic minority students with failure, and some instructors went so far as to cite "ESL issues" specifically as a challenge for the student. From our perspective, however, while most instructors interpreted English-language problems to mean a language or language variety other than Standard American English, we viewed English-language problems much more broadly to mean mechanics and syntax (when such characteristics affect meaning in writing), characteristics that could have easily been applicable to U.S. domestic students. As I am a WPA, it goes without saying that no instructor should ever fail a student simply because of English-language problems, and no single instructor cited only that reason for the student failing. With reference to language, academic discourse, and retention, Pegeen Reichert Powell argues,

> What we're really talking about when we're talking about the exclusionary practices of academic discourse and the rules of Standard American English that writing teachers are expected to enforce is retention, the question of whether or not students will persist once they are in. My point here is that the discourse of retention must qualify, even radically revise, our arguments about improving access. (16)

For us, the purpose of including this question on our questionnaire was not to enforce the retention of students who have met instructors' expectations of proficiency in Standard American English or academic discourse. Instead, our purpose was to determine the individual student's needs that our one-on-one faculty facilitators would have to address for each student. For future semesters of this program, we have decided to frame the question differently, while also informing instructors of the fact that we aren't simply referring to students with language and language variety differences.

That said, we couldn't help but notice the disproportionate number and increase of U.S. racial minority and international student referrals that were linked to English-language problems. We became concerned that some instructors might be racially profiling students in such a way that shifts responsibility for teaching these students away from the instructor. In this way instructors have created a present-day division of labor that shifts the responsibility from current ENGL 1050 instructors to rescue section intensive instructors, who might provide additional teaching and support. In "Composition Studies and ESL Writing: A Disciplinary Division of Labor," Paul Kei Matsuda reminds us that "one of the consequences of the disciplinary division of labor is the lack of concern about the needs of ESL writers among composition specialists that continues even today" (714). Further, because most of our ENGL 1050 instructors do not have specialty training in the teaching of ESL or students who speak other varieties of English, many interpreted the shifting of such students to the rescue sections as an opportunity to remove students from their classes. At least one instructor inquired about the rescue sections as an opportunity to remove ESL students during the first week of the spring 2015 semester before any single student could possibly provide evidence of failure! Indeed, framing language in relation to both deficit and failure continues to bear negative consequences for our college campuses, particularly for multilingual writers. For me, changing these policies and consequences begins with WPAs.

But They Cheat Their Asses Off! Cultural and Language Biases and Plagiarism

The second initiative that I will discuss in this chapter concerns my campus's treatment of L2 and ESL learners in the teaching of writing across the university. After having many conversations with various constituents across campus, the writing center director and I decided to apply for a racial healing grant, cosponsored by the Kellogg Foundation. This grant was established after an external consultant was hired to conduct a diversity climate survey of

WMU's campus. Some of the findings from the diversity climate study point to the need for additional support in relation to equity and inclusion. As such, we decided to apply for this grant to support ESL and L2 learners because we were fed up with culturally and racially biased assumptions about what L2 and ESL writers can and cannot do and their misconstrued propensity to plagiarize. We recounted all of the horror stories we had heard from faculty. One particular story that comes to mind pertains to a faculty member who told us "those Chinese students cheat their asses off!" After submitting our application, we were approved to study ESL and L2 students who were being reported for plagiarism and to bring in national scholars of ESL writing and plagiarism in composition studies to provide professional development training to instructors on teaching about plagiarism across campus. We are currently in the data collection phase for ESL and L2 students. The initial step in our effort to bring about change in the WMU understanding of plagiarism and why students plagiarize will consist of a multipart assessment that will serve inclusion by providing data about (1) WMU student writers' plagiarism insights and practices; (2) WMU instructors' understanding of best practices in helping all students, but especially L2 students, engage in writing without plagiarizing; and (3) WMU instructors' varied practices for dealing with perceived incidents of student plagiarism, which are not currently consistent, although our university does have an established academic integrity policy and related procedure in place.

Researchers in composition studies and TESOL (teaching English to speakers of other languages) agree that many students may struggle to learn to write without plagiarizing and to engage with their learning through original research writing projects (Pecorari; Ramanathan and Atkinson; Moss; Duff). However, these researchers also agree that for L2 students, achieving such learning can be challenging because the notion of plagiarism operating in U.S. universities—which undergirds students' learning from writing assignments, assessment of writing assignments, and students' grades on such assessments—is a dominant academic cultural construction; however, it is not often a part of the cultural and school backgrounds of students in the groups of ESL and L2 (Pecorari).

In addition to scholarship cited previously, our grant proposal also drew on what Muriel Harris recognizes as "local research" (3). Harris explains that because writing center staffs engage in numerous confidential and safe discussions with diverse students about the students' writing assignments and needs, writing centers have a unique, broad view of campus writing pedagogy concerns and should share their insights about campus writing trends (their "local research") with their institution's community. Accordingly, writing center local insights about student plagiarism at our university affirm what writing studies and TESOL researchers say about students who plagiarize: it is far more common for L2 students to need help understanding how to write without plagiarizing than it is for students who are more fluent in English writing and more academically experienced with the U.S. conventions of academic citation. In fact, our 2013–14 writing center records indicate that 157 different students had consultations in which unintentional plagiarism was discussed. Each of those clients was either an L2 or academically underprepared student.

When applying for our grant, we described in our narrative the context of how many of our consultants witnessed and documented the struggles that L2 students bring to their plagiarized text. Our narrative described how consultants have noted how a restricted English vocabulary often limits the ability of L2 students to paraphrase and to understand what needs to be cited and how consultants have helped L2 students grapple with cultural concepts and writing strategies that many WMU instructors assume all students already know or should know. Consultants have seen how international and domestic L2 students are often confused by the cultural understandings of idea "ownership," individual work, and the notion that source citing is a professional practice rather than a disrespectful suggestion that instructors may not know the sources. Consultants have also noted specific writing concerns, such as that L2 students may focus on the goal of providing correct text to instructors rather than on the goal of sharing original versions of ideas they have gathered from research, and consultants know that some L2 students use computer translation devices to translate passages from scholarship in their first language and will include the translations in their papers without citations because they see no reason to provide instructors with sources the instructors cannot read. Following English writing strategies they learned in non-WMU TESOL programs, some L2 writing center clients "paraphrase" by changing every third or fourth word (excluding articles), while some have their English-speaking friends edit their papers so that the native English speakers can paraphrase the text without mimicking the original too closely. But then the papers that the L2 students submit do not represent their true ability, and some pay for too much writing assistance from others.

As a WPA, then, SRTOL must become a critical component for how we teach students about plagiarism. At our campus, we notice that international students, ethnic minority students, and even students of lower socioeconomic backgrounds are the ones most often being reported by instructors for plagiarism. Specifically, as a WPA, I've found that instructors who reported such students say that they easily detected plagiarism because the writer's voice and language differed dramatically from the language of the texts from which the writer supposedly plagiarized. One example that illustrates this move immediately comes to mind. When one of our speakers, funded by the racial healing grant, came to visit, the speaker sat in on my graduate TA training course, ENGL 6690: Methods in Teaching College Writing. For this particular session, I asked ENGL 1050 TAs to bring up any recent instances of plagiarism they may have encountered. One inexperienced instructor noted one case where she was watching the student for supposed plagiarism. She—and I'm not really paraphrasing here—said that she was surprised that her student's previous writing assignments were written well because "you know, he's a minority," in this case, African American. When it came to the assignment for which he supposedly plagiarized, the instructor wanted to retroactively grade his previous work because he may have plagiarized on those assignments as well; after all, they were well written without many surface-level errors. It took everything for me to bite my tongue. Fortunately, the visiting speaker, with a different subject position from me, stepped in and demonstrated a tactfully teachable-moment exercise for the TA.

Unfortunately, the TA's assumption that a racial minority student had plagiarized in an essay because the language closely resembled Standard American English is quite common. Countless studies reveal racist assumptions about international students, particularly those who come from racial minority backgrounds, and their false propensity to plagiarize. In the article "'White Pages' in the Academy: Plagiarism, Consumption and Racist Rationalities," Sue Saltmarsh writes, "Discourses of racialised 'Others' of the education market, together with the discursive visibility associated particularly with language use, it is argued, disproportionately mark out international students as those primarily associated with illicit practices such as plagiarism." When speaking directly about the relationships between plagiarism and institutional racism, Saltmarsh argues that educators who accuse ethnic minorities of plagiarism often

> shift the emphasis away from pathologising and Othering discourses, and query those cultural assumptions and biases that impede, rather than enhance, the learning outcomes for these growing student cohorts. In particular, such arguments speak to the extent to which "epistemological racism" (Cadman, 2005, p. 1) is an inherent feature of institutional structures of power, through which consumers of tertiary education are attracted, progress, and are occasionally excluded.

In the two examples I discussed in this section, "epistemological racism" was imbedded in institutional systems of power and authority. In both cases, instructors who work within the authoritative structures of institutions made negative cultural and racist assumptions about the writing and ethical practices of racial minority students. In the first example, an instructor made the assumption that Chinese students cheat; and in the second example, the TA assumed that an African American student must have plagiarized because his writing too closely resembled Standard American English. This TA made negative assumptions about what the writer could and could not do. Instead of interpreting the writer's use of Standard American English as an example of "codemeshing," where the writer treats Standard American English and African American Language "as part of a single integrated system" (Canagarajah, "Codemeshing" 403), the instructor assumed that the student did not possess the skills necessary to write multilingual and translingual texts. She further suggested that he did not possess this ability "because you know, he's a minority," thus reflecting racial biases. For me, it is clear that racist and cultural biases not only exist in how we interpret purported plagiarized texts by students, but they also exist in whom we accuse and penalize for plagiarism.

Assessment and Support: Resistance and Refusal with our ENGL 1050 L2 and ESL Research Project

As part of our third initiative, the ENGL 1050 assessment project for L2 writing (also aimed at retention efforts to assist L2 and ESL students in their first-year writing courses), we proposed to study L2 and ESL ENGL 1050 students to determine the needs, resources, and support such learners might need. We sought to recruit students and code their papers for linguistic (syntactical, lexical, morphological) and rhetorical patterns common across texts; to hold focus groups with ENGL 1050 L2 students about needs and desired resources; and to hold focus groups with instructors teaching L2 students in their ENGL 1050 sections. After receiving funding, I personally experienced direct resistance from ENGL 1050 instructors when I explicitly asked, as the WPA, for permission to recruit students from their sections, although these same instructors were often eager and more than willing to report L2 writers for instances of plagiarism or to refer those with failing grades to the spring 2015 rescue sections of ENGL 1050.[3]

As WPA I was surprised at the outright refusal to refer L2 writers to our study. Many instructors even claimed they did not have any L2 writers enrolled in any of their sections despite the fact that, at the time of the study, there were more than 125 visa-issued or permanent-resident students enrolled in ENGL 1050 whose countries of origin did not include English as the primary or an official language spoken.[4] Students whose profiles fit these characteristics come from countries such as Saudi Arabia, China, the United Arab Emirates, Brazil, and Argentina. This would suggest that if international students from such countries of origin make up at least 5 percent of our total population of ENGL 1050 students, we should indeed have more referrals. Of course, this does not include total numbers of U.S. domestic students whose first language is other than English, nor does it automatically suggest that students from these backgrounds indeed did not learn English as a first language. My point is simply that there is an audience and population for our study despite the claims and resistance from instructors that this audience does not exist. Most instructors poll their students with preliminary course questionnaires that ask them if they learned English as a first, second, or subsequent language, so instructors are aware of what L2 and ESL students are in their courses.

Similar to the issues we encountered when trying to recruit L2 students from instructors' classes, we have also experienced resistance from instructors to making referrals to our rescue sections, despite the fact that, unlike our L2 writing research study, all instructors are required to participate and notify me, director of the First-Year Writing Program, whether or not they have students to refer. As WPA I've had to send out three or four reminders to various instructors who refused to respond. Of those who did respond, many referred L2 students who they determined would fail, but they would not recommend these same students to our L2 writing study. At first, during the fall 2014 semester, some instructors refused to refer students, particularly first-generation, working-class, and ethnic minority students, because they were unwilling to endorse those students' potential to respond to intensive instruction, and they only began to refer such students during the spring 2015 semester after we instructed them that we would determine whether or not the students were potential candidates based on their application. On my campus, it goes without notice that the resistance I have experienced pertains to my attempt to provide support for racial minority students and L2 writers, many of whom are also ethnically minority.

Perhaps resistance by instructors—in particular, early-career and/or novice instructors—also comes from their failing to recognize implications for retention initiatives at their institution, and in higher education more generally, beyond their very localized classrooms; they therefore do not pay enough attention to the role of first-year writing and its population of first-year students in the institution where they teach. But, as Powell reminds us, "faculty who regularly teach first-year students, such as writing instructors, should especially pay attention to the discourse about retention at their respective institutions because a fairly common assumption in retention scholarship is that efforts should be focused on first-year students" (7). The fact that most of our ENGL 1050 instructors are either part-time or TAs is significant because these groups of instructors often see themselves as those who are least likely to effect any sort of institutional change and, therefore, are often resistant to buying into campus-wide initiatives that lend themselves to promoting or effecting change. Further, since instructors are limited in the compensation they receive, they often see any such initiatives as additional uncompensated work, even though our study did offer some incentives for their time and participation.

Implications

My experiences with the three initiatives described in this chapter each point to the need to understand implications for SRTOL in relation to racial, cultural, and linguistic prejudices. Writing instruction across our colleges and universities continues to operate from a theoretical assumption of deficit. Students who come from particular racialized and linguistic backgrounds are assumed to require remediation when instructors have decided that they cannot and will not teach such students. Even the code language of our rescue sections of ENGL 1050 assumed that such students need to be rescued. As I noted previously, we have since revised our rescue section courses and titled them ENGL 1050 intensive courses, although such identification suggests that students are cases or patients in need of intensive care. The question then becomes, at what point do we create a celebratory discourse around students who come from particular racial and linguistic backgrounds? At what point do our writing programs promote philosophies that honor and celebrate the richness of the writing practices that such students bring with them? At one point our discipline, as Suresh Canagarajah has suggested, merely tolerated such students using their home language varieties "for informal purposes and low-stakes writing needs" ("Place" 602). For me, however, it seems that many of our institutions—mine included—aren't tolerating such students very well at all in most rhetorical situations; as I stated earlier in this chapter, one instructor asked that a student be removed from her class completely! What will happen to our discipline, and our position statements crafted so vigorously by the CCCC Language Policy Committee, if trends continue to move students backward?

It also goes without saying that my subject position, as an African American untenured WPA who led and collaborated in these initiatives (I have now achieved tenure!), also plays a critical part in the resistance I have experienced from instructors, most of whom are TAs who have been coached to resist my authority by faculty mentors (implications of subjectivity and authority are another essay). Arnetha F. Ball reminds us in "Expanding the Dialogue on Culture as a Critical Component When Assessing Writing":

> It is time to include the voices of teachers [and WPAs] from diverse backgrounds in discussions concerning writing assessment. And as we do so, we will find that these voices have much to add that cannot only inform, but re-shape current assessment practices, research priorities, and policy debates that focus on finding solutions to the challenges that face us as we try to improve writing assessment for a diverse population. (380)

For me, then, I see kairotic occasion for composition to consider the CCCC's SRTOL as an opportunity to use assessment to work toward social justice for speakers and writers of languages and language varieties other than Standard American English, and I see the inclusion of people who look like me who are WPAs as possessing the potential to contribute and do this sort of work.

The final critical implication I will mention pertains to understanding applications of SRTOL on our college campus in relation to instances of plagiarism. Notice that in a couple of the examples I shared, including one from my ENGL 6690 class, an instructor identified plagiarism because there was significant distance between the student's home language and the academic language the student sought to appropriate. The question then becomes, however, what happens to native English speakers who plagiarize when their native language closely resembles the academic language. Are they off the hook? Perhaps an SRTOL application might suggest that students not only have the right to their own language but, also, the right not to have their language used against them for cases such as plagiarism. In short, students whose languages or language varieties differ from academic English should not be disproportionately reported in cases of campus plagiarism. An SRTOL policy that promotes equality in how we identify and punish students guilty of plagiarism is essential to promoting institutional changes in diversity, equity, and inclusion by broadening faculty and student awareness of poststructuralist, multimedia plagiarism theories of writing pedagogy. Essential too is awareness of assessment strategies that can help all students learn to understand the U.S. plagiarism construct, avoid unintentional and intentional plagiarism, and engage in their writing development.

Appendix A: ENGL 1050 Intensive—Student Application

This form can be emailed to Staci.Perryman-Clark@wmich.edu or submitted to an administrative assistant, 6th floor, Sprau Tower

Name: Instructor Name:

What issues do you think you had in English 1050? (circle all that apply)

    Absences
    Missing papers
    Late work
    Missing homework
    Poor quality of work
    Assignments that did not meet criteria
    Failure to work through the writing process
    English language problems
    Other (explain)

In a short paragraph, please explain why you think you would be successful in an Intensive section:

Appendix B: ENGL 1050 Intensive—Instructor Recommendation

Instructor Name: Student:

Why is this student failing English 1050? (circle all that apply)

    Absences   Number of times absent:
    Missing papers   Number of missing papers:
    Late work
    Missing homework
    Poor quality of work
    Assignments that did not meet criteria
    Failure to work through the writing process
    English language problems
    Other (explain)

In one or two sentences, please explain why you think this student would work more successfully in an Intensive section:

Notes

1. The CCCC first affirmed and published the resolution in 1974; since then the CCCC has reaffirmed it twice. The full text of the statement is available here: http://cccc.ncte.org/cccc/resources/positions/srtolsummary.

2. In *The Practice of Everyday Life*, Michel de Certeau distinguishes between *strategies* and *tactics*. He associates strategies with institutions, those systems of power and authority. Individuals, on the other hand, are associated with tactics because

they are not afforded the power that is most frequently connected to institutional infrastructures or systems. Because of my marked subject position as both an African American female and as an individual, I am required to use tactics and not strategies to fight oppressive practices of institutional systems like higher education.

3. Although one-fifth, or six of the thirty students, who enrolled in our rescue sections classified themselves as L2 learners, many more ESL and L2 students were referred to our rescue sections by instructors, especially during the spring semester. These students, however, did not fill out an application.

4. The statistics were obtained from enrollment records accessible through our university's Banner system.

Works Cited

Ball, Arnetha F. "Expanding the Dialogue on Culture as a Critical Component When Assessing Writing." *Assessing Writing*, vol. 4, no. 2, 1997, pp. 169–202. Reprinted in *Assessing Writing: A Critical Sourcebook*, edited by Brian Huot and Peggy O'Neill, Bedford/St. Martin's / National Council of Teachers of English, 2009, pp. 357–86.

Canagarajah, Suresh. "Codemeshing in Academic Writing: Identifying Teachable Strategies of Translanguaging." *The Modern Language Journal*, vol. 95, no. 3, 2011, pp. 401–17.

———. "The Place of World Englishes in Composition: Pluralization Continued." *College Composition and Communication*, vol. 57, no. 4, 2006, pp. 586–619.

Duff, Patricia A. "Language Socialization into Academic Discourse Communities." *Annual Review of Applied Linguistics*, vol. 30, 2010, pp. 169–92.

Harris, Muriel. "Diverse Research Methodologies at Work for Diverse Audiences: Shaping the Writing Center to the Institution." *The Writing Program Administrator as Researcher: Inquiry in Action and Reflection*, edited by Shirley K. Rose and Irwin Weiser, Boynton/Cook Heinemann, 1999, pp. 1–17.

LaHaie, Jeanne and Adrienne Redding. "1050 Intensive First-Year Writing Program Data Analysis (2014–2015)." College of Arts and Sciences Deans and Chairs Council, Western Michigan University, 15 June 2015.

Matsuda, Paul Kei. "Composition Studies and ESL Writing: A Disciplinary Division of Labor." *College Composition and Communication*, vol. 50, no. 4, 1999, pp. 699–721.

Moss, Beverly J. *A Community Text Arises: A Literate Text and a Literacy Tradition in African-American Churches*. Hampton Press, 2003.

Pecorari, Diane. "The Use and Misuse of Reference Sources: Plagiarism or Patchwriting?" BALEAP Professional Issues Meeting on Academic Writing, 27 Nov. 1999, Reading, UK. *CityU Scholars*, scholars.cityu.edu.hk/en/publications/publication(309aff99-4050-43de-a032-bb39f4da4e30).html.

Perryman-Clark, Staci M. *Afrocentric Teacher-Research: Rethinking Appropriateness and Inclusion*. Peter Lang, 2013.

Perryman-Clark, Staci, et al., editors. *Students' Right to Their Own Language: A Critical Sourcebook*. Bedford/St. Martin's / National Council of Teachers of English, 2015.

Powell, Pegeen Reichert. *Retention and Resistance: Writing Instruction and the Students Who Leave*. UP of Colorado, 2013.

Ramanathan, Vai, and Dwight Atkinson. "Individualism, Academic Writing, and ESL Writers." *Journal of Second Language Writing*, vol. 8, no. 1, 1999, pp. 45–75.

Richardson, Elaine. *African American Literacies*. Routledge, 2003.

Richardson, Elaine, and Keith Gilyard. "Students' Right to Possibility: Basic Writing and African American Rhetoric." *Insurrections: Approaches to Resistance in Composition Studies*, edited by Andrea Greenbaum, State U of New York P, 2001, pp. 37–51.

Saltmarsh, Sue. "'White Pages' in the Academy: Plagiarism, Consumption and Racist Rationalities." *International Journal for Educational Integrity*, vol. 1, no. 1, 2005, doi:http://dx.doi.org/10.21913/IJEI.v1i1.17. PDF unpaginated.

Smitherman, Geneva. "'The Blacker the Berry, the Sweeter the Juice': African American Student Writers and the National Assessment of Educational Progress." National Council of Teachers of English Annual Convention, Nov. 1993, Pittsburgh, PA.

Section IV

Digital Perspectives

10

What Do Humans Do Best?
Developing Communicative Humans in the
Changing Socio-Cyborgian Landscape

Charles Bazerman

Our writing programs are a result of our pasts and a response to current pressures, needs, and opportunities, as explored in a number of the essays in this volume. The work of these programs, however, is addressed to the future—the lives our students will live in their future worlds. Their careers may have them writing well past 2060, and if they are teachers, some of their students will be writing into the twenty-second century. This essay looks at current trends and dynamics to make judgments about what writing skills will be useful tomorrow and the day after tomorrow, as technology takes over many tasks, freeing us to do what humans do best.

Technology ever increasingly is taking over the work previously done by humans in the composition, distribution, storage, access, and use of communications, and is doing new tasks previously unimagined. What will the human half of the cyborg need to be able to do? Cyborgs fascinate us with technological extensions overcoming human limitations—powerful exoskeletons, infrared vision, and real-time data scans, but we forget the human half (see Bazerman, "WAC"). Every technological extension requires new training, orientations, sense making, and decisions by the living, educated, and purposeful legacy neuro-biomass—us and our students.

Until recently, legacy has been considered a valuable gift, associated power, explicit in the etymological root *legate*. To understand the legacy that humans bring as we move forward, we must look backward to appreciate human value and power. Academic, professional, commercial, and personal writing have been from their beginnings sociocommunicative endeavors mediated by symbolic and material technologies. Humans created the richness of written language out of odd practices of making marks on clay. Using these marks, humans have created documents to coordinate meanings, knowledge, thoughts, plans, and activities. Humans have invented vocabularies, spelling systems, representational means, genres, forms of organization and persuasion, kinds of knowledge, channels of communication and distribution, procedures and locations for archiving and access, and all the other things skilled writers have to be aware of and make choices about.

Through much of history, the balance of the work of composing marks to convey meanings rested on individuals working alone or in small collaborative groups, even as text reproduction, distribution, and archiving extended networks of collaborators. Writers struggled with the changing language and expanding representational means, relying on the cultural inheritance from the work of previous humans—as resources and structuring frameworks for their current work. As we invent new technologies, new social arrangements and endeavors, new forms human-machine collaboration, new representational and distributional means, and new storage and analytic devices, the work equation is changing, shifting what each human or collaborative group of humans is expected to do and creatively add at any particular moment.

This changing distribution of work means that human skills also must change. While machines will come to do what machines do best, humans must reallocate their attention and skills to do what humans do best in these socio-cyborgian activity systems. Further, humans need to develop new skills to understand, direct, and make choices about these complex networks. The teaching of academic, technical, and professional writing just a few decades ago focused on individuals producing fairly stable forms, but now the field has refocused on social participation and collaboration within organizations through changing genres and changing situations, incorporating new technologies for document production and distribution. Nonetheless, technical and professional writing has still focused on the individual text as a discrete utterance and the center of attention, even as the text has come to include hypertext and multimedia, and even as the work of production has included new possibilities of collaboration and distribution. If we are to address the needs of communicators entering careers that will span through the middle of the twenty-first century, we need to make further shifts in our perspective so that we can identify what humans can best contribute to social-individual human-machine complexes and then prepare our students to do those things better. Further, we must prepare our students to understand and maintain executive control of these rapidly evolving communicative systems.

The Transformation of Inscription Professions

To see what these shifts of the work equation might entail, let us look at a few related inscription professions that have been even more rapidly affected by digital technologies. In accounting, the traditional roles of keeping and then inspecting the financial records of individuals and enterprises have transformed as organizations grew, requiring ways to manage internal flows of material and labor resources, to minimize costs and maximize efficiencies. Records needed to be coordinated and aggregated across multiple sites, even as management became centralized in head offices. Communicative and informational technologies, from the memo and form to the filing cabinet and graphic charts of business trends, created new tasks and roles for white-collar workers and managers. Massive records formed the neural infrastructure of corporate intelligence. While those at the top of the organizational pyramid gained more information in aggregated and often quantitative form to make creative managerial decisions, those lower down were often routinized through highly typified communications. Keeping records of clients and services as enterprises grew created new challenges to keep order, make sense, and predict future business (Yates, *Control*). The insurance industry with its need for actuarial prediction and cost models that extended over decades, along with its many individualized consumers having individual contracts, payment, and expenditure records, was at the forefront of the accounting revolution and formed the clients for IBM, as insurers were inventing information technologies even before electronic computing (Yates, *Structuring*). Accounting was already turning into executive oversight and prediction based on cost and income models. Comparison of alternative ways of doing business, reflected in data representations, created the opportunity for higher-level choice making, leading to the Taylorist transformation of business.

When digital technologies, and in particular the VisiCalc and Excel spreadsheet programs came along in the late 1970s and early 1980s (Ceruzzi and Grad), accounting underwent a further transformation. The traditional skills of fine handwriting, orderly columns, and accurate and rapid calculation were displaced by the technology. Accountants and managers now needed to be able to use and understand the operations carried out by the software, so as to make choices of how to represent items within the spreadsheet and which calculations and re-representations they wanted the spreadsheet to carry out. Human energy was freed to compare alternative scenarios under different assumptions or with different ways of representing costs and assets. The work of accounting and management became ever more one of manipulating numbers and alternatives to maximize profits. The world of financial markets became more complex and intense, with rapid decisions made on massive data projecting shorter time periods, relying on rapid calculations done by machine, and with actions often triggered by machines (though according to human-set parameters). At the same time, the data reported and calculated on became more abstracted and distanced from actual material properties, labor, and other assets, as all were transformed into numbers and calculations buried within machine operations. This often led to sleight-of-hand vanishing of information necessary for accurate concrete evaluation, as notoriously happened with credit default swaps and other manipulations of the financial crisis. Even without deceptive manipulation, evaluation and decision-making about assets require special skills and tools.

Additionally, as billing and financial-transfer work becomes distributed through various electronic means, new authentication, security, and oversight challenges require the design of new accounting systems and inspection of transactions. Not only have internal calculation and oversight of transactions been taken over by machines but so has the production of final reports in standardized formats. Here is a simple example that exhibits some of the changes that have occurred. Many years ago, when I first used a tax accountant, what impressed me were his neat handwriting, precise columns, and rapid calculation, along with his knowledge of tax law and IRS procedures. My current tax accountant still knows the law and IRS procedures, but he hardly writes any numbers by hand and makes few calculations. He does, however, understand the tax software installed on his computer, and he knows how to test the consequences of reporting income and expenses under one category or another. Much of our meeting time is devoted to trying out different scenarios, reporting the information under various categories and adopting different assumptions and parameters. Then the machine calculates and prints out the completed forms and a cover letter telling me exactly what I owe, how to mail it, when and how much I should pay in estimated taxes for the next year, and so on. The machine speaks to me in intelligible boilerplate including the specifics of my own finances and obligations. But once I approve and sign these documents, the relevant data (and not the whole form) are submitted to the IRS electronically through an authenticated system.

I could continue with many other examples. In the European Union, where the increasing cooperation across borders has amplified the need for government and business documents to appear in multiple languages, various forms of machine translations, specialized real-corpus concordance tools, and other databases have transformed the work of translators, who are as likely to spend their time overseeing and making choices about options offered by their digital tools as they are seeking words and phrases in their mind to express meanings presented in the original language and then transcribing them (Alcina). Literary translation has not yet been as influenced by technological changes, but the world of gaming, with its highly typified language use, does allow people to participate through multiple languages. So the work of linguists and translators moves to another level, designing systems to aid translation and creating interactive environments within which language and interactions are structured to allow multiple-language equivalents. The translator steps out of the individual interaction to the level of systems design or inspection and correction of machine-produced alternatives, with particular attention paid to pragmatics, idiom, and interaction.

The Technologizing of Writing

So let's now think about writing: What work involved in writing is being off-loaded onto technology, and what tasks remain for humans? In some ways writing is a clunky technology, requiring many years to become reasonably competent at and often consuming much time to compose individual texts. Written texts are frequently cognitively demanding for the reader, inviting interpretation, approximation, and imprecision of meaning, even misunderstanding. Further, the effectiveness of communication through writing relies on a state of trust and receptivity between writer and reader, for the reader must be willing to mentally construct meaning initiated by another mind (Bazerman, *Rhetoric*).

The inscription of letters or characters is the first clunkiness that people learning to write encounter, whether with a stylus forming cuneiform on clay, a brush forming ideographic characters on scrolls, or a pencil forming alphabetic letters in school notebooks. Much of writing education over millennia has been devoted to teaching fine motor control and visual discrimination, manipulation of writing instruments, form and decipherment of characters, spelling, arrangement of symbols on the medium, and so on. In every child's life, five or more years are devoted to gaining reasonable competence in transcribing words and sentences. Technology has long been easing those burdens, replacing stylus and brush with pens and pencils of increasing ease and reliability, and simplifying letterforms and scripts. Over the last century and a half, the easier motor task of pressing keys has been displacing burdensome handwriting. Electronic devices have further eased neat and accurate transcription by preformatted page layouts and spell checkers. Variable sizing of fonts and displays eases development of visual discrimination. Voice recognition and transcription from voice to text and text to voice are also becoming increasingly common—especially for those with disabilities that interfere with transcription and visual recognition. But we still teach versions of hand-transcription skills and spelling along with keyboarding because they are still necessary in many circumstances and because they appear to be useful for becoming familiar and skillful with written language (James and Engelhardt). Even as we off-load tasks to machines, we still need understanding and skills to make choices, to oversee processes, and to monitor results so that we retain ultimate responsibility for the production.

Word choice, grammar, syntax, and text organization similarly require long training, extending through secondary and higher education, to create texts that are intelligible to others and indicate the education and thoughtfulness of the writer—and thus the reliability of the meanings evoked by the texts. While we have become used to employing digitally implemented prostheses that help guide and correct us, from online dictionaries and thesauruses to grammar checkers and document templates, these have longer print histories that still require us to make choices about their suggestions.

Revision as well has been a burdensome task requiring use of expensive material and time resources, hard-to-decipher text markups, repeated copying of versions, and production of clean copies. The introduction of computers and visual displays was immediately recognized as a great boon to revision, decreasing the amount of manual labor and cost in making revisions and trying alternative formulations (Hawisher). Nonetheless we remain responsible for the final choices. Reproduction and distribution of texts have also become increasingly easy and inexpensive, aided by the evolution of the printing press, paper technology, rail and motor transport, and now desktop publishing and internet distribution and cloud storage in multiple formats and synchronicities. Yet we are still faced with choices about privacy and distribution in various networks.

Bibliographic technologies from library catalogues and journal indexes to Google and citation searches have also aided the amassing of intertextual resources and positioning our writing intertextually. Distant reading now helps us assess large corpora of texts to discover patterns and aggregate tendencies, aiding us in our own new statements in relation to the intertexts of interest (Moretti). Even cutting and pasting have made quotation (and plagiarism) easier, and automated summarizing may soon be coming to your desktop (Lehmam). Even the existence of Wikipedia as a ready source indicates how technologically assisted crowdsourcing diminishes the work of individuals in locating and elaborating information, even while increasing the challenges of choice making (Silva). Certainly such large-corpus crowd-based strategies have made machine translation more intelligible and less comic. Already some kinds of reports that are highly typified and need only specific data are regularly produced. Even the first press-release reports of earthquakes and volcanic events are machine produced in order to speed distribution of the information (Oremus).

Monitoring and Higher-Level Decision-Making

As changing supportive technologies have sped the process and decreased the burden of communication, the work of writers has moved from basic transcription and word memory to determining larger purposes, making design choices, selecting content, adopting analytic perspectives, hybridizing options, and monitoring and refining results to match our intention and the needs of the communicative situation. The lower-level tasks that continue to be taken over by machines have each been considered at various times the very definition of writing and the focus of writing education—starting with letter or character inscription; then continuing with spelling, syntactical, and grammatical correctness; and most recently focusing on text organization, content location and selection, and intertextual context and positioning. These latter (which are currently only partially supported technologically) are still central to contemporary writing pedagogy. Further, all of them, even transcription, require some continuing instruction for the purposes of understanding, monitoring, correcting, and refining machine choices. Yet we writers are also increasingly freed from their more burdensome aspects to engage in higher levels of understanding, decision, and control—previously either the work of specialized experts or unimagined. Typographic and page design was until recently the work of only a few print and book design specialists. While syntactic and grammatical correctness still remains an educational obsession, the expectations for compact, efficient, and pointed sentences continue to rise among educated writers. That is, the specialized work of identifying and tightening good jokes and tight lines is extending beyond the work of professional comics and poets. Even the technological constraints of Twitter inspire some to epigrammatic and allusive elegance.

Even skills now considered higher order are getting technological support. Idea-mapping software helps aggregate and organize ideas, resources, and bibliography, bringing the challenging tasks of complex text organization to higher levels of coherence and creativity. While basic PowerPoint at first seemed to pull us back to traditional outlining and five-paragraph speeches that some of us worked hard to get beyond, it has in practice facilitated the use of graphics, illustrations, videos, and sound files (along with sidelong comments) into the linearity of our arguments. Further, visual displays can carry audiences through more subtle forms of organization, making the steps of reasoning, focal points, memorable slogans, and stances more transparent. Other less linear presentation software, such as Prezi, offers new ways of displaying coherence and movement through ideas by graphic means. So even while we still need to monitor the neatness, spelling, and phrasing of the texts we insert, we are able to develop whole new sets of skills to make more complex and higher-order decisions. So this monitoring and higher-level decision-making is the first thing we already do (and we already support in our teaching), but we will need to do it more, with greater scope for incorporating, organizing, and monitoring the products of nonhuman entities.

Collaboration and Collective Production

The second thing we already do well, but will be called on to do in more complex ways facilitated by technology, is collaboration, as we move more fully into a distributed and crowdsourced world with new temporalities, interactivity, and social arrangements. It has long been known that books are collaborative, a collaboration that Martha Woodmansee recovered from a 1753 German economic lexicon:

> Many people work on this ware [the book] before it is complete and becomes an actual book in this sense. The scholar and writer, the papermaker, the type founder, the typesetter and the printer, the proofreader, the publisher, the book-binder, sometimes even the gilder and the brass-worker, etc. Thus many mouths are fed by this branch of manufacture. (15–16)

Scientific publication has required increasingly large numbers of participants (Clarke), and this tendency is growing throughout academic writing. Collaboration has been facilitated by technology, and it is not uncommon for small groups of collaborators distributed around the world to work on texts together, even simultaneously, and reports of organizations can move from one department and one work group to another with immense speed and around the clock. The text and work space can exist in the cloud with participants visiting and interacting without even having to hand the documents back and forth. Software facilitates group revision, tracking who makes what changes when, who approves, and what the current state of the document is.

The complex collaborations facilitated by technology require ever-greater project management skills to keep people coordinated and on task and to maintain textual coherence and purpose. Maintaining timely advance of the work and managing the synchronicities and asynchronicities, attention, and timely input and decision-making by participants out of physical reach—all require special skills that we need to develop (Orlikowski and Yates).

The distributed writing supported by recent technologies, however, goes far beyond the enhanced management and negotiation of traditional collaboration in order to engage crowdsourced negotiation of knowledge among large numbers of strangers and large collections of knowledge from many sources. Sometimes these projects seem to grow on their own, with viral creativity. Yet there is always some organizing intelligence in the initial concept and the design and maintenance of the platform. Somebody or somebodies have to have the idea that Twitters or Vines or Tumblrs would be a good idea; to design a workable platform that allows the activity to flourish; to monitor the products to make improvements; and to publicize and instigate use. Almost all require medium-to-large companies to keep them going, comply with government regulation, and make them financially viable. Wikipedia, perhaps the ultimate crowdsourced project, has had to create many technologically supported layers of peer and accredited-expert monitoring and editing.

Here is an example of the complexity of skills that go into writing such a project, though we might not traditionally think of these as writing skills. I have recently contributed to the *Naming What We Know* project organized by Linda Adler-Kassner and Elizabeth Wardle. In order to gain the collective view of the field, Linda and Liz instigated crowdsourcing, interaction, and negotiation by a group of field leaders to identify key threshold concepts. They invited almost fifty scholars in the field to participate in a wiki. Once the participants were aligned to the goals of the project and the idea of threshold concepts, they each entered, on a wiki, brief paragraph descriptions of up to three candidate threshold concepts of composition. As the descriptions appeared, participants were asked to comment on each other's offerings. Debate ensued about what was meant, how the concepts might be better formulated, what was the relationship to others' proposals, whether these were true threshold concepts or just ideas about writing, and so on. In order to keep the scholars' attention and engagement, the schedule was tight, with only about a month for the initial gathering and then another month for discussion.

While the editors hoped that a few of the concepts would rapidly rise as obvious choices, in fact the editors needed to sort through the ideas, as not every one was presented in an appropriate form, many overlapped, the discussion did not resolve the boundaries and contradictions, and a number of other incoherencies obscured clarity. So the editors, communicating with each other by email phone and Dropbox, reformulated, combined, and reorganized the candidate proposals, while trying to stay true to the original expressions. The editors then posted the new hierarchically organized list on the wiki and asked for further discussion and comments.

The editors then commissioned participants, based on their interests in the previous discussion, to write more extended discussions of the concepts. Again, the deadline for production of the chapters was quick in coming, about two months. As chapters emerged, issues of consistency of audience, tone, stance, level of handling the subject, and citation density had to be worked out—through peer commentary and revision comments from the editors. A second wiki was established for the emerging full text, to provide space for peer commentary, revision, boundary definition, and cross-referencing. This process of mutual adjustment and alignment continued on the new wiki and through email correspondence between editors and authors as well as between authors of different chapters At this stage comments from authors about topics that were not fully handled led to some new topics and rearrangement of sections, and last-minute authorships were assigned. At this point the editing of this section of the book became more like conventional editing of a reference work, with the editors corresponding with the authors about final revisions.

Some of the skills exhibited by the editors, such as controlling attention, urgency, and temporality in asynchronous virtual exchanges or creating negotiations among stakeholders to come to shared statements, have been discussed in the literature, but I don't think we have yet gotten a good conceptual grasp on the skills exhibited here of identifying untapped energies and providing them a shape, managing hybrid crowdsourced projects, knowing when to give participants a long leash and when to rein them in, and knowing when to take charge. The more we can articulate and reflect on these skills, the better the already skilled will become at them, the more widely shared the skills will be, and the better we can prepare our students for writing in distributed, asynchronous, knowledge- and project-building networks. While technology can aid this work, it is the human understanding of how to bring humans together for complex collaborations that directs the use of technologies to facilitate shared writing.

Audience and Empathy

The third thing humans are already good at is knowing audiences and forming positive relations with them. This is hardly news to teachers of writing, informed by wisdom going back to Aristotle and almost every rhetoric and rhetorician since then. So here I will add only a few comments on technological approaches to audience. Appealing to audiences is a subtle skill, and mechanistic demographic approaches soon become transparent and alienating. Just think about how we react to the personalized selection of ads we get based on the data gathered from our Google searches and web visits. As Plato knew, speaking to someone else's soul requires an intimate sense of the other person, gained most easily and fully in face-to-face dialog, requiring a spontaneity, authenticity, and intuitive sense of the other that leads to surprise and poignant intervention in individuals' thinking and action. Demographic approaches to audience used by current technologies rely on our most visible characteristics and behavior, and do not empathetically reconstruct our ways of feeling and thinking. Rather, they are in constant danger of evoking audience responses of feeling stereotyped or pigeonholed in ways that do not understand who the persons addressed perceive themselves to be or how they feel.

In writing we have to enact that relationship at a distance of time and space, an act that requires even greater human understanding and intuition. Writing to audiences is an act of imagination facilitated by knowledge of genres and activity systems but still one requiring a particularity in the specific communication (Bazerman, *Rhetoric* and *Theory*). If imagining oneself into the place and relations of writing in the paper-and-ink world is an imaginative challenge, the affordances of technology for distribution and access make the imaginative challenges of the digital world even greater. The kinds of disasters that people regularly get into by wayward emails or retweets going viral are only the spectacular tip of the much deeper set of challenges that await us.

The human potential for intuitive alignment with each other (a potential possibly shared with other higher forms of life) may have to do with mirror neurons. Mirror neurons were discovered about twenty-five years ago in monkeys (di Pellegrino et al.) and then in humans (Keysers and Gazzola). They are thought to allow one organism to interpret or sense the observed behavior of another organism as though it were enacted within the observer's own neural system, so, for example, when we watch another human (or even a horse) racing, our own adrenaline starts to pump. These mirror neurons, whose function and consequences are still controversial, may be crucial to such phenomena as understanding intentions, motor mimicry, vocal imitation, acting, empathy, self-awareness, and even language (Keysers). Mirror neurons may be central to how humans reenact meaning and affect from texts as well as project anticipated reactions of readers. If mirror neurons are crucial in anticipating audience relation and response, then the human ability to address audiences is unlikely to be matched by technology for the foreseeable future. Even if somehow technology were to simulate mirror neurons, the receiving entity would still have to have a rich set of humanlike experiences and a humanlike capacity for sense making in order to reverberate in alignment with the anticipated audiences.

Intuition and Emergent Cultural Consciousness

The final human capacity I want to explore has also to do with intuitive response to our social situation and the role of imagination, inspiration, or the muse—how fresh thoughts come to our mind. At one level, seeding imagination is a familiar part of our pedagogy as well as part of our personal tricks for invoking the muse by creating good environments for work and concentration, by engaging in stimulating discussions and brainstorming, by using such practices as freewriting and meditation. Early in my career as a writer and teacher I encountered the words of one of the great chess masters, which I remember as something like "study the position, calculate some alternatives but then sit quietly until one of the moves arises in your mind with conviction." When I am writing, that is what I try to do: set the conditions, set my goals, understand the direction and probable end point of my argument, consider some alternatives about the next phrase or sentence or paragraph or section, or even the theme and title of an essay—and then I wait until the right words arise in my mind. I take the risk of following my intuition, buffered by the knowledge that I can always revise, go back to a false turn, erase an aberration. But the more experience I have, the less often I find I have to retrace or erase my steps. That intuitive assessment of where next to go seems to factor in all I know about the subject, situation, and emerging text. I have come to recognize and calibrate the rising certainty within me and follow it.

When words and ideas flow, I am sometimes surprised by what emerges, but not shocked, because in retrospect this is where my understanding of the situation was leading me, and the words that appear suggest a way to go through the opening that was prepared before. As writing teachers, we repeat to students quotations about not knowing what you think until you write it. Most writing teachers also diminish the risk of trusting intuitions, through techniques like brainstorming, freewriting, or discovery drafts.

What might this curious role of intuition in writing mean about humans and how we operate? Of course we can always invoke a spark from the gods or muses or genius. But cognitive neuroscience may give us some clues about what is going on. When I first heard thirty years ago about neural networks and connectionism, they seemed to provide a mechanism for the way our minds make complex intuitive judgments and assessments below the level of conscious calculation, through the weight of experience, perception, disposition, and goals. Then, in the 1990s, work by Damasio and others identified emotions as overall assessments of situations that are central to the process of making judgments and shaping actions—not in ways that were inimical to reason, but at the very core of reason. Damasio's more recent work gives further speculative insight into the nature of consciousness and its relation to more primitive, emotive parts of the brain. I need to confess, though, that I have a very limited window into the world of neural science, working primarily from Damasio's popularizing books: *Descartes' Error, The Feeling of What Happens, In Search of Spinoza,* and *Self Comes to Mind.* Further, although his work depends on the work of many others, I know that work largely through his representations, and not all of his colleagues are likely to agree with his syntheses and assessments.

Damasio presents a view of consciousness as a mapping and awareness of one's situation and a calculation of alternatives, which rests on nonconscious mappings and also lags behind knowledge, calculation, and assessment. Damasio points out that, in the most basic of animals, the neural system records sensations from the external senses, and the creature uses those neural representations to locate itself and move in the environment without necessitating consciousness. As organisms become more complex, they record and map information from internal sources as well to regulate their internal state and operations, such as with peristalsis, heartbeat, and balance. Again, these do not require consciousness. As organisms become neurologically even more complex, however, they map their neurological system's operations and responses, forming the bases of an emergent consciousness. A creature's mapping of its neurological state forms a sense of a protoself, which gives rise to an awareness of a self that is having sensations and emotions and moving or acting in response—what Damasio calls an autobiographical self. Yet conscious feeling or awareness lags behind the initiating emotions and more basic neurological responses. One begins to smile before one recognizes one is smiling; one begins releasing the fight or flight chemicals before one recognizes feeling fearful or threatened. Nonetheless, the conscious awareness allows one to monitor oneself and change one's action in relation to other things one notices or feels. One can consciously suppress the smile or choose to ignore a pain in order to finish a race, and one can proceed observantly and intentionally without fleeing or fighting when in a threatening situation. Thus, in neurologically complex creatures, there are multiple levels of mapping one's perceived external and internal environment, one's responses to the states of the environment, one's self-directed goals that require monitoring multiple impulses, and ultimately one's choice-making processes. While higher-order parts of the frontal cortex are involved in this mapping of the self, the more anterior regions of the brainstem where emotions emanate take a central role in integrating information and setting the orientation and directions of consciousness. With the emergence of an autobiographical self that feels and acts and is aware of feelings and actions over time, one's experiences accumulate and modify one's internal state, influencing future feelings and actions.

Yet this knowledge and personal development are locked within one's neural system and, without communication, die when the neural system dies. Means of communication, particularly language, however, allow one to share one's perception of the world, one's internal states, and one's choices and actions, to the degree that one can formulate and represent them. With language, knowledge becomes shared, and the loneliness of the separate neural system begins to be breached. One's idiosyncratic representations, perceptions, and knowledge can also be compared with those of others, and one's errors or bad judgments can be modified with goals and choices changed on the basis of the shared communal mappings or knowledge. We can speculate beyond the scope of Damasio's examination of the internal state and consciousness to characterize language itself as facilitating a communal mapping of the world and coordinating the internal states of the people sharing language, plans, and procedures—thereby forming community knowledge. This community knowledge then changes the perceptions, action environment, and goals for each of the individuals, as humans perceive the world in relation to and as part of the social collective. Language, as many have noted, embodies the collective unconscious, embodying motives, perceptions, dispositions, and orientations that are not yet brought into explicit collective consciousness. Community elders, seers, poets and other creative writers, essayists, philosophers, and other visionaries can serve to articulate those underlying social impulses and emotions as part of an emerging collective consciousness. Again, each new level of consciousness lags behind the emergent representations, orientations, and actions of individuals and the collective. Yet each new level of emergent representation and eventual consciousness embodies and is built on the mappings of prior states of knowledge and awareness.

Literacy and inscribed knowledge creates a new level of more enduring and more widely spread representations, more widely testable against broader experience, and even explicitly tested against probes into the environing world as well as the internal states and actions of oneself and one's fellow creatures. Each extension of the formation and diffusion helps advance the sophistication and reliability of the communally shared representation, in the manner in which Eisenstein identifies the impact of the printing press in the formation of science (Eisenstein). Yet even with all the sophistication of the literate mappings and communal knowledge, which form the basis of our posing the problems of living and all the subproblems of inquiry and action choices, calculative awareness lags behind the internal intuitive assessment that directs our action and sense of where answers and actions lie. We only become aware of these sensed solutions as they arise to consciousness, even while they embody all the prior layers of conscious learning and sharing of knowledge. So even as we are puzzling through the conflicting representations of multiple works of philosophy and science, based on our lifetime of experience, reading, debate, and writing, we gravitate toward our solutions only partly aware of where we are headed and what formulations will make sense to us. We are like the reptile wandering toward the warmth of the sun and rocks but only recognizing where we are as we get there. We work best when we do not fight those inarticulate emotions arising into feelings of where we are headed and then honoring them by bringing them into consciousness. Humans are really good at intuition, and it is going to be a long time, if ever, before machines will beat us at this game. Computers now can beat us at chess, but only because they can calculate all the alternatives much faster in this closed domain with discrete moves conducive to calculation. Even within this realm, some humans, with far less calculative power but lots of intuitive understanding and access to the collective experience of play embodied in the extensive literature on chess, can still do pretty well. Most of what we write about is far less bound and far less conducive to a limited set of calculations than chess. In those more open domains of life, human intuition, emotions, and subconscious calculation still are powerful guides to statements we articulate for our conscious recognition and to contend for a place within communal consciousness.

Conclusion

Writing and knowledge technologies offer us a wealth of cyborgian extensions. Yet as we and our students learn to take advantage of them, we also need to learn to take more explicit advantage of our human capacities within the cyborgian partnership. Here I have proposed some things that humans do well, and might need to do even better, to use our technology to best advantage: monitoring production and making high-level choices; coordinating complex distributed collaborative projects; understanding audiences empathically and forming communicative relations; and, finally, sensing where we are and where we are headed in complex social-informative-knowledge worlds. Or we could say in a more familiar way: revision, rhetoric, collaboration, inspiration.

A wag once said the future will be much the same, only different. Or perhaps the future will be different, but much the same. The conditions and situations of writing will be different, but we will continue to do well what we do well, while technology does what it does even better.

Works Cited

Adler-Kassner, Linda, and Elizabeth Wardle. *Naming What We Know*. UP of Colorado, 2015.

Alcina, Amparo. "Translation Technologies: Scope, Tools and Resources." *Target*, vol. 20, no. 1, 2008, pp. 79–102.

Bazerman, Charles. *A Rhetoric of Literate Action*. Parlor Press / Writing across the Curriculum Clearinghouse, 2013.

———. *A Theory of Literate Action*. Parlor Press / Writing across the Curriculum Clearinghouse, 2013.

———. "WAC for Cyborgs: Discursive Thought in Information Rich Environments." *Labor, Writing Technologies, and the Shaping of Composition in the Academy*, edited by Pam Takayoshi and Patricia Sullivan, Hampton Press, 2007, pp. 97–110.

Ceruzzi, Paul, and Burton Grad, editors. *Spreadsheets for Everyone*. Special issue of *IEEE Annals of the History of Computing*, vol. 29, no. 3, 2007.

Clarke, Beverly L. "Multiple Authorship Trends in Scientific Papers." *Science*, vol. 143, no. 3608, 21 Feb. 1964, pp. 822–24.

Damasio, Antonio. *Descartes' Error: Emotion, Reason, and the Human Brain*. Putnam, 1994.

———. *The Feeling of What Happens: Body and Emotion in the Making of Consciousness*. Harcourt, 1999.

———. *Looking for Spinoza: Joy, Sorrow, and the Feeling Brain*. Harcourt, 2003.

———. *Self Comes to Mind: Constructing the Conscious Brain*. Pantheon, 2010.

di Pellegrino, Giuseppe, et al. "Understanding Motor Events: A Neurophysiological Study." *Experimental Brain Research*, vol. 91, 1992, pp. 176–80.

Eisenstein, Elizabeth L. *The Printing Press as an Agent of Change: Communications and Cultural Transformations in Early-Modern Europe*. Cambridge UP, 1979.

Hawisher, Gail E. "The Effects of Word Processing on the Revision Strategies of College Freshmen." *Research in the Teaching of English*, vol. 21, no. 2, 1987, pp. 145–59.

James, Karin H., and Laura Engelhardt. "The Effects of Handwriting Experience on Functional Brain Development in Pre-literate Children." *Trends in Neuroscience and Education*, vol. 1, 2012, pp. 32–42.

Keysers, Christian. *The Empathic Brain*. Social Brain Press, 2011.

Keysers, Christian, and Valeria Gazzola. "Social Neuroscience: Mirror Neurons Recorded in Humans." *Current Biology*, vol. 20, no. 8, 2010, pp. 353–54.

Lehmam, Abderrafih. "Essential Summarizer: Innovative Automatic Text Summarization Software in Twenty Languages." *Proceedings, RIAO '10: Adaptivity, Personalization and Fusion of Heterogeneous Information*, Apr. 2010, CID, Paris, pp. 216–17.

Moretti, Frank. *Distant Reading*. Verso Books, 2013.

Oremus, Will. "The First News Report on the L.A. Earthquake Was Written by a Robot." *Future Tense, Slate* blogs, 17 Mar. 2014, www.slate.com/blogs/future_tense/2014/03/17/quakebot_los_angeles_times_robot_journalist_writes_article_on_la_earthquake.html.

Orlikowski, Wanda J., and Joanne Yates. "It's About Time: Temporal Structuring in Organizations." *Organization Science*, vol. 13, no. 6, 2002, pp. 684–700.

Silva, Mary Lourdes. "Can I Google That? A Case Study of the Online Navigational Literacy and Information Literacy Strategies of Undergraduate Students in a Research-Writing Course. *The New Digital Scholar: Exploring and Enriching the Research and Writing Practices of NextGen Students*, edited by Randall McClure and James Purdy, Information Today Inc., 2013, pp. 161–88.

Woodmansee, Martha. "On the Authorship Effect: Recovering Collectivity." *The Construction of Authorship: Textual Appropriation in Law and Literature*, edited by Woodmansee and Peter Jaszi, Duke UP, 1994, pp. 15–28.

Yates, Joanne. *Control through Communication: The Rise of System in American Management*. Johns Hopkins UP, 1989.

———. *Structuring the Information Age: Life Insurance and Technology in the Twentieth Century*. Johns Hopkins UP, 2005.

11

From *Topoi* to Tweets: What Ancient Rhetoricians Can Teach Digital Natives

James A. Herrick

Writing is here to stay, despite ancient reservations about its effect on memory and present-day worries over its potential degradation by new technologies. While we cannot ignore developments affecting expression, or changes in the skills of students, it is also important to keep in view the stunning plasticity of rhetoric—the discipline that provides writing its historical foundation and technical resources. This chapter explores several lessons for the contemporary teacher and practitioner of writing that are embedded in the very earliest efforts to formulate a theory and practice of rhetoric. Interpreting truncated messages, forging connections with disparate audiences, and discovering resources for making words do practical work—all are quandaries of today's authors addressed by ancient rhetoricians. Rhetoric endures as the practice of effective public expression.

"I call it foul, as I do all ugly things." With this caustic summary, Socrates—a mouthpiece for Plato's views in the dialogue *Gorgias*—dismisses rhetoric as beneath the dignity of a legitimate philosopher. Rhetoric was a means of flattering audiences, misleading the citizenry with emotional appeals and hypnotic sounds, and perverting public opinion on the most important issues, particularly justice. In so characterizing rhetoric, Plato marked the discipline to the present day. About thirty years after Plato inscribed his disparaging comments, the great philosopher's most famous pupil, Aristotle, presented to his students a series of afternoon lectures on rhetoric. Student notes from these lectures form the basis for Aristotle's *Rhetoric*, a book that remains the most influential and among the most perceptive treatments of the art ever produced.

I want to take this brief introduction to the conflicting responses elicited by the ancient art of rhetoric as a point of entry into the discussion of an enormously important question: How can writing, and allied skills such as critical thinking, be taught successfully to a generation of digital natives,[1] those born after the creation of the brave new online world? I was, until a few years ago, a disgruntled traditionalist, regularly and with gusto expressing my suspicions about how the internet was undermining writing skills, corrupting student research, shortening attention spans, and generally ruining both education and democracy. However, I have since decided that there is no real pedagogical future in lamenting the fact that college students text, tweet, blog, and post pretty much all day and all night long, often blissfully unconcerned that their writing may not be up to traditional standards.

This was the world my digital-native students inhabited, and it seemed to me that the challenge I faced as a teacher was to discover an approach to teaching written and spoken expression and criticism that was flexible, practical, and grounded in real-world language practices. But, wasn't this a description of the very *techne* of logos I had been studying for decades, the robust, ever-adapting discipline of strategic language use practiced by Protagoras and Gorgias, Isocrates and Aristotle, Cicero and Augustine? Why reinvent the discursive wheel when such a well-tested conveyance was so close at hand and so many skilled pilots could be so readily summoned?

Sticking with my vehicular metaphor, allow me to back up just a bit. Let's take a look at the Mediterranean world in which rhetoric was born. A renaissance of sorts was under way in the Greek polis. As classicist Michael Gagarin has written,

> The second half of the fifth century was a period of intellectual innovation throughout the Greek world, nowhere more so than in Athens. Poets, philosophers, medical writers and practitioners, religious reformers, historians, and others introduced new ways of thinking. (6)

The art of rhetoric was an enormous success in the Greek-speaking world of the fifth and fourth centuries. Moreover, rhetoric emerged as a popular study during an era of profound and rapid social change, not unlike our own.[2]

Athens was undergoing a shift away from aristocratic rule and toward democracy. As historian of rhetoric John Poulakos writes, "When the Sophists appeared on the horizon of the Hellenic city-states, they found themselves in the midst of an enormous cultural change" ("Terms" 56). Rhetoricians such as Corax, Protagoras, Lysias, Gorgias, Hippias, and Isocrates offered training in inventing arguments and presenting them in a persuasive manner to a large audience. Newly enfranchised citizens in Athens and other city-states created a market for something not previously available in Greece, a form of higher education centered on the effective use of language and reasoning in public as well as private settings. I say private, because the rhetoricians also promised to teach the proper management of one's *oikos*, or household. So, rhetoric was from the beginning a highly adaptable art of effective expression and practical judgment.[3]

Rhetoric fomented something of a cultural revolution in the Greek polis: the key to personal success and influence was no longer class but was skill in persuasive speaking. The study and practice of rhetoric had a far greater impact on ancient Athens than did now famous philosophers such as Plato. "Plato's influence on fourth-century Athenian culture was relatively slight," writes Gagarin, "whereas oratory was central to the lives of most Athenian citizens, who regularly attended meetings of the courts or the Assembly in some capacity . . ." (5).

That is some of the historical background for ancient rhetoric. On a more personal note, I have always been impressed with one thing in particular about the ancient rhetoricians, including the Greek and Sicilian sophists who have received such bad press but who actually invented the discipline of rhetoric. They were aware of, and great advocates for, the power of spoken and written language, and they recognized that skill with language afforded the skilled rhetorician access to power. This is probably why the aristocratic Plato hated them so much.

The sophists were practitioners and teachers of the force and beauty of well-crafted *logoi*: words, arguments, language, particularly as deployed before large audiences. Rhetoric equipped advocates to employ language, not just effectively, but with aesthetic force and expressive elegance. For Gorgias, words did not merely *represent*; they were a living force that shaped thought and experience. The ancients saw rhetoric as power, not simply as a utilitarian speechmaking method. Jane Tompkins has noted that "the equation of language with power, characteristic of Greek at least from the time of Gorgias the rhetorician, explains the enormous energies devoted to the study of rhetoric in the ancient world" (203–04). The great Roman orator Cicero was not a member of an elite family with automatic access to power. He soared to political prominence on the wings of his remarkable rhetorical skill. The same kind of thing happens to this day, which is one reason I am a proponent of teaching and indeed of requiring courses in critical thinking, rhetorical analysis, and rhetorical theory.

I should also note that the ancient rhetoricians were language *innovators*; they experimented with the technology of strategic language, recognizing in the science of persuasion a transformative, almost magical capacity. The rhetoricians found persuasive language a flexible, adaptable medium; for some it was the most powerful force in the human world. Ancient rhetoricians attributed the founding of civilization itself to the cooperative activity made possible through oratory. Rhetoric was closely aligned with magic; its stylistic devices were more than ornaments—they were means of capturing and focusing the power of verbal expression. Carefully crafted language shapes thought and moves others to think and act. The ancient rhetoricians offer surprising insights into the power of words and the important social role played by persuasion. These early students of language, according to Gagarin, loved to challenge "traditional ways of thinking" (16).

Language was creative magic, a means of shaping the world anew (de Romilly, *Magic*). Moreover, this power of strategically crafted words might be made available to anyone. The rhetoricians celebrated the clash of arguments— the *dissoi logoi*—as a method of rational investigation rather than as evidence of a failure to grasp the truth. These ancient values may have discovered a new home in the contemporary digital world. It seems that with all of the enthusiastic blogging, tweeting, and web page construction going on today, the newer digital formats have revived for digital natives some of these very ideas: language as a powerful, malleable, liberating, even enchanting force.

Rhetoric for the Residents of Digitopia

The digital experience is a powerful and permanent feature of contemporary life. Google processes 100 billion searches each month. Twitter averages 500 million tweets a day, about 6,000 per second. The record was an astonishing 143,000 in a single second. Facebook has 2.19 billion active monthly users worldwide, with 80 percent outside the United States. It goes without saying that young people today will face a job market radically altered by digitization, a political arena in which Twitter and Facebook are potent forms of political expression and organization, and an incomprehensibly powerful internet. Computer scientist Hugo de Garis predicts an internet that is 1 trillion times more powerful by 2050 (de Garis). As futurist Ray Kurzweil is fond of reminding audiences, technology continues to grow exponentially in its capacity and sophistication, but that also means it grows in its influence on how we live our lives. What sort of preparation should educational institutions be providing in response to these profound changes? I believe the question should be phrased a little differently: What kind of people are we hoping our students will become in the face of these massive cultural changes?

Here's one answer. A research team at one University of California campus completed a major study of internet use by students. According to the study's lead researcher, the team discovered that "spending time online is essential for young people to pick up the social and technical skills they need to be competent citizens in the digital age" ("Study"). This phrase caught my attention: "competent citizens in the digital age." I asked myself, Is *this* what I'm hoping for as the kind of person my students will become? The UC team did not choose a nuanced and interesting phrase like "flourishing individuals who understand how to live successfully in a digital age" but chose the oddly Orwellian slogan, "competent citizens in the digital age."

This reflection took me back to a rhetorician from whom I have learned some valuable lessons, Richard Lanham. Lanham has also pondered the question of how to structure a curriculum of higher education today. Some time ago he called for a return to rhetorical studies as a way of preparing students to understand the impact of computers on how we read, write, and think. Rather than developing a completely new theory of literacy for the computer age, which some have proposed, Lanham argued in the 1980s, at the beginning of the digital age, that "we need to go back to the original Western thinking about reading and writing—the rhetorical *paideia* that provided the backbone of Western education for 2,000 years" (Lanham, *Electronic Word* 51). For Lanham, the highly adaptable study that originally taught the Western world its approach to public communication, in fact its way of thinking, can still teach us new things (Lanham, "Rhetorical Paideia").

What was Lanham noticing in the ancient study of rhetoric? The rhetorician Protagoras boasted that he could teach his students to make the weaker case appear the stronger, a claim that concerned many traditional Athenians. Was this idle sophistic bombast, or was Protagoras perhaps making a claim about a practical, rational skill that is as relevant in the digital age as it was in ancient Greece?[4] Several valuable skills, as it turns out. First, the student of rhetoric gained rational agility. Rhetoric taught the capacity to make sense of a jumble of details, to distinguish meaningful fact from mere assertion, and to arrive at a reasonable judgment accordingly—a virtue Aristotle labeled *phronesis*. Second, rhetoric developed awareness of *kairos*, sensitivity to situational and cultural nuances, an intuition regarding the discursive possibilities inherent in a setting, a topic, and the civic atmosphere, all converging in a moment of decision. Third, rhetoric taught a nuanced appreciation of audiences. The ancient rhetoricians were relentlessly audience focused; every phrase, trope, and argument had instantly to connect with the rational practices, aesthetic preferences, and heartfelt beliefs (*doxa*) of a living, breathing audience of ordinary citizens. Every argument was an enthymeme constructed jointly by audience and speaker.

These aspects of rhetoric—its capacity to inculcate practical judgment, situational awareness, and an appreciation for audiences—are the keys to the art's perennial utility. The study of classical rhetoric is thus ideally suited to equipping students to adapt to the ever-shifting demands of today's information explosion, constantly changing cultural landscape, and diverse audiences.

Embarked on the search for a new model of literacy, Professor Andrea Lunsford headed up the Stanford Study of Writing, which examined nearly fifteen thousand undergraduate writing samples (Stanford Study). Lunsford arrived at an intriguing conclusion after considering so many instances of how and what students today write: "I think we're in the midst of a literacy revolution the likes of which we haven't seen since Greek civilization" (qtd. in Thompson, "Clive Thompson"). According to Lunsford, a "new literacy" is emerging as digital formats such as blogs and personal web pages encourage *more* writing, not less. Moreover, the writing that the student generation participates in is active, public, strategic, and persuasive—in other words, traditionally rhetorical. *Wired* magazine reported that "Lunsford's team found that the students were remarkably adept at what rhetoricians call *kairos*—assessing their audience and adapting their tone and technique to best get their point across" (Thompson, "Clive Thompson"). It certainly got my attention that Lunsford selected the foundational rhetorical term *kairos* to describe her team's observations about student writing, virtually reviving in one linguistic maneuver the entire classical rhetorical tradition for the digital age.

Kairos directed the skilled advocate's attention to an opportune moment of decision-making. It allowed the orator to sense an opening for judgment, like the momentary breach a skilled weaver exploits when the shuttle is passed through the threads of the warp. Lunsford finds student writing today to be "conversational and public, which makes it closer to the Greek tradition of argument" than to forms such as the traditional letter or book ("Clive Thompson"). Conversational, public, yes, but also kairotic—charged with the opportunity for a decision, an opportunity arising out of the public clash of arguments, the interaction of ideas, the constructive connection between a speaker and an audience.

We may be witnessing, in the digital writing revolution that Lunsford is tracking, a return to such traditional rhetorical concerns as kairos, phronesis, enthymeme, and doxa. We hear these concerns reflected in the Stanford Study's finding that students "defined good prose as something that had an effect on the world. For them, writing is about persuading and organizing and debating . . ." ("Clive Thompson"). These are qualities of discourse that would have been recognized, and taught, in Athens more than 2,400 years ago. Lunsford suggests that the digital domain is, for many of today's students, a location for pursuing vibrant public discourse, not just entertainment, personal communication, or quick information. That is to say, rhetoric can be practiced well online.

I would like to focus attention on four aspects of the digital experience that, despite the many advantages the internet provides, pose challenges to the free and constructive exchange of ideas and about which the ancient rhetoricians can provide us some guidance. I have titled the sections covering these four concerns "Managing Information," "Addressing Abbreviation," "Assisting Investigation," and "Public Participation."

Managing Information

One of the conditions we face in the digital age is an inundation of information. Where information is abundant, skills of organization and interpretation are indispensable, capacities that contributed to the practical wisdom of phronesis. The art of rhetoric cultivated the capacity to make sense of information, form it into a meaningful narrative, render it comprehensible to an audience, and employ it as a guide to rational civic decision-making. The volume of data created over the entire history of the human race will be doubled in the next four years. Now, I am not suggesting that any one of us, regardless of rhetorical training, is equal to the task of making sense of such staggering volumes of data. But, neither am I willing to say we should throw up our hands in desperation, only to be swept away by the digital deluge. So, what can be done to cultivate phronesis under such conditions?

A great deal of information is not in and of itself an aid to better thinking, better living, or improved democracy. The role of human agents as interpreters of that information may get lost in the data deluge. Not until bits of data are prioritized and interpreted through an interaction with personal and cultural values do they contribute to any of the goals listed above.

Rhetoric as a discipline has, from its inception, made matters of evidence and its interpretation a central concern. The rhetorician, practicing the virtue of phronesis, was an observer and interpreter of evidence, a mediator between an audience of citizens and a world of complex and often conflicting messages. The rhetor's goal was to interpret, in the public sphere, evidence from various sources—to transform inarticulate data into plausible narratives that addressed issues of public concern. Aristotle noted early in *Rhetoric* that rhetoric instructs one in how to adapt complex messages to an audience of ordinary listeners, listeners who could not quickly take in a long and complex argument. Rhetoric's unrelenting focus on the audience and doxa required orators to consider the accessibility and relevance of their evidence. Locating, prioritizing, and interpreting relevant information will be increasingly critical as the digital age matures and are thus central concerns of teachers of writing and critical thinking.

It is well known that for Aristotle, rhetoric was "the faculty of observing in any situation the available means of persuasion." To develop the virtue of phronesis in my students, I encourage them to be observers and interpreters rather than simply retrievers and reporters. To observe is to perceive with the intention of finding meaning in a scene. I observe not to answer the question "What is here?" but to answer the question "What is going on here?"

Training in rhetoric should involve instruction in critical observation. The topical, or *loci*, systems are relevant here; *topoi* were catalogues of distinct and recurring forms of argument that were aids to invention. That is to say, the topoi provided a frame for interpreting evidence. Evidence does not speak for itself; it has to be interpreted by human agents guided by personal values and tested ways of reasoning. Topical systems reflected trusted ways of reasoning, but also ways of turning and testing evidence in the process of invention.

It can be helpful to provide students with assignments that cause them to observe and then to interpret evidence. I ask my argumentation students to explore a Gallup Poll opinion survey on the topic of public attitudes toward Charles Darwin and evolution. They observe the findings and then respond to just two questions: First, is there any finding that surprised you? Second, is there any finding that particularly interested you? We then work on questions of interpretation. For example, I ask them to think about what it tells us that 44 percent of American adults cannot associate the name Charles Darwin with any scientific theory. We then discuss what I take to be topical issues such as the possible causal explanations of such a finding, the presence of analogous cases in public understanding or misunderstanding of science, and the pragmatic consequences for society of such a situation. This kind of exercise emphasizes the students' role as observers and interpreters of data and also as framers of arguments.

In a more advanced assignment I ask students to read broadly on a controversial topic of their choosing, directing them to observe facts and arguments that strike them as particularly intriguing or persuasive. I encourage them to imagine how these might be interpreted using various topoi such as reasoning by cause, by analogy, by generalization, by example, or by principle. They then write a brief persuasive essay in which they develop and deploy three of these lines of argument in support of a thesis on the topic. The students supply support for their arguments from evidence discovered through their reading, but the emphasis throughout remains on the lines of argument they have invented. The goal of this assignment is to help students become intentional observers, careful interpreters of the evidence landscape, and empowered reasoners. I take these qualities to be characteristics of phronesis.

In another exercise I provide students with a series of findings that they are to treat as evidence. An example finding would be that the United States incarcerates more of its citizens than any other country, about 2.3 million. I then ask them to draw a conclusion from the finding. Their conclusions range from "policies that result in mass incarceration need to be changed" to "the police in the United States are particularly effective in apprehending criminals." I then ask students to suggest one or two personal values that they believe connect the finding to the conclusion they derived from it. This last step emphasizes the presence of doxa, or personal beliefs and values, that guide the interpretation of evidence. Students often are surprised by the range of conclusions at which they and their classmates have arrived, which increases their appreciation for the diversity of belief present in audiences as they interpret evidence.

In addition to instruction in topoi, the ancient discipline of rhetoric helped students to manage evidence by cultivating a capacity for narrative. For instance, *narratio*, or a narration of events, was a crucial component of a judicial speech in Rome. The Greek ceremonial, or *epideictic*, speech explored and refined public values by means of a narrative—a showing forth—focused on praise or blame.

Narrative continues to play a crucial role in organizing and interpreting evidence. British philosopher of science Mary Midgley has observed that "facts will never appear to us as brute and meaningless. They will always organize themselves into some sort of story, some drama." Midgley notes that when the narratives of science go unexamined, "these dramas can indeed be dangerous" (4). I have introduced students to narratives and metaphors that shape public perceptions of science and technology, arenas they often assume to be free of such rhetorical devices. They are struck by how powerfully these narratives influence their own understanding of technological developments.

In another exercise my rhetorical criticism students read Barack Obama's first inaugural address. Among the questions I ask them to consider is one about the president's use of narrative structures in his speech, particularly a narrative of journeying together as a nation. Students can see how the president's development of evidence in the speech follows the contours of his journeying narrative. The discussion cultivates an appreciation for narrative as one means of translating information for an audience or of transforming data into cultural wisdom. If, as Midgley suggests, facts organize themselves along the arc of a narrative, then comprehending the force of narrative takes on great urgency as a component of phronesis in an era of burgeoning data.

Addressing Abbreviation

Skills of evidence interpretation and awareness of both audience values and the structure of arguments are crucial in addressing another fact of our digital existence—message abbreviation. Recently I have been attempting to reconstitute complete or nearly complete arguments from severely abbreviated messages found in tweets, guided by Aristotle's concept of the enthymeme as discussed in his *Rhetoric*. Aristotle recognized that public arguments are not like the logician's formal syllogism—they don't have as many parts when publicly articulated. This absence of some vital information introduces greater likelihood of ambiguity and misinterpretation, though it may also enhance persuasion. In any event, messages online are often brief, even cryptic, and thus pose challenges to writer and audience alike.

Twitter has emerged as both a quick means of communication and an important forum for public discourse. Twitter tends to obscure the rigid speaker-audience relationship that characterizes more traditional argument formats such as the public speech or debate. Moreover, the enforced constraint of 280 characters requires extraordinary economy of expression; effective tweets require careful word choice, a direct writing style, and abbreviation. Such enforced economy can render writing more engaging and vibrant. At the same time the role of abbreviated communication in civic discourse and particularly decision-making may be compromised by abbreviation. In such a context the audience must supply more of the message. This is where what I have come to think of as enthymematic reading may be of assistance.

Here's an example I have used in a teaching exercise. Not long ago Canada's supreme court struck down legal barriers to prostitution, a decision prompting thousands of messages on Twitter. One tweet read, "I would like to thank the Supreme Court of Canada for declaring sex workers to be persons." This may be—as the author suggests—simply a note of thanks, but it sounds to me like there is lurking here an embedded argument.

The writer suggests a connection between the Canadian court's decision and the issue of personhood. What belief or value would connect the concepts? I might try this: when one's occupation is recognized as legal by government institutions, one's personhood is affirmed. Admittedly, this is a guess, but the tweet suggests that it is close to what was intended by the writer.

The complete argument I am suggesting reads like this:

1. Canada's supreme court struck down laws against prostitution, effectively declaring prostitution legal.

2. When one's occupation is recognized as legal by government institutions, one's personhood is affirmed.
3. Thus, the Canadian supreme court has declared sex workers to be persons.

This argument remains enthymematic—it is shaped around a vital connection between the shared convictions, or doxa, of the writer and an intended audience. This kind of exercise prompts students to think more critically, not just about what they are reading, but also about what they have left out of their own messages and, in so doing, have left their audience to supply. Students again ask themselves about their own values and whether their assumptions are fair and reasonable. Aristotle's understanding of the intimate cognitive and emotional connection between authors and audiences, his "enthymeme insight," emerges as relevant to digital natives inhabiting a rhetorical realm often characterized by abbreviation.

Assisting Investigation

Closely related to skills of interpretation are skills of investigation. Ancient rhetoricians employed the method of dialectic in their teaching, requiring students to invent arguments for and against a proposition. This approach taught students to argue either side of a case and thus to develop investigative skills that might reveal insights hidden in an easily dismissed minority position—the weaker case. I find the approach friendly to a pluralistic and diverse cultural setting, which the digital world certainly is, as it teaches one to explore a range of points of view. In dialectical instruction, argument met counterargument in a series of exchanges that might, at the outset, cast doubt on the apparently stronger case but that eventually yielded a more nuanced view of the truth. Dialectical method was rooted in the notion of dissoi logoi, or the clash of arguments for or against a claim. Aristotle agreed with the sophists that rhetoric taught advocates to reason on both sides of an issue and that this practice was investigative in nature.

Maryanne Wolf of Tufts University has said, "We are not *what* we read, we are *how* we read" (qtd. in Carr, "Google"). Her concern is that the online experience teaches superficial ways of encountering texts, inculcating reading habits that substitute "efficiency" and "immediacy" for deeper analysis. The ancient rhetoricians' teaching methods trained students to analyze cases critically, to think with agility, to ask probing questions, to pose counterarguments, and, yes, to respond eloquently. Susan Jarratt and Rory Ong write, "rhetorical training created a critical climate within which to question, analyze, and imagine differences in group thought and action" (16). Such a critical climate invites nuanced and inquisitive thought while at the same time returns us to the aesthetic considerations inherent in kairos.

Kairos acknowledged that good judgment depended on a careful consideration of audience, arguments, circumstances, evidence, and the need for a reasonable and timely decision. Kairos also connoted verbal decorum, a concern for speaking words appropriate to the situation, the issue, and the audience. Thus, kairos emerged at the confluence of pragmatic, situational, and stylistic considerations—the right word, spoken at the right moment, and in the right manner. Isocrates writes in his *Panathenaicus* that rhetorically educated persons are those "who manage well the circumstances which they encounter day by day, and who possess a judgment which is accurate in meeting occasions as they arise and rarely misses the expedient course of action" (sec. 30). In a digital context that flattens expression, it is important to recall that kairos implied more than simply clear judgment about a course of action; it also implied clear judgment about a mode of expression that met the occasion. As investigation turns to decision-making, the advocate guided by kairos speaks at the opportune moment with precision and force but also with an aesthetic awareness that connects rigorous investigation with elegant expression.

Public Participation

Nicholas Carr argues that an ecology of interruption characterizes the online experience, an environment of structured distractions designed to ensure that we not spend too much time on any one page or any single activity. Such designed distraction is rooted in the economics of the digital realm—the more pages we visit, the more advertising we experience (*Shallows* 104). Linda Stone coined the term "continuous partial attention" to describe the mental frame of many of us today—never fully focused on any one message, attention always at least partially scattered (Stone). This experience of distraction can also cause us to lose our sense of involvement in a broader world, a public.

For Protagoras, no proposition had been established until it withstood the scrutiny of counterargument. Presenting and defending a claim requires paying attention, not just to the claim itself, but to its opposite. John Poulakos writes that Protagoras "provides a worldview with rhetoric at its center," and this rhetorical worldview "demands of the human subject a multiple awareness, an awareness at once cognizant of its own position and those positions opposing it" ("Terms" 60). Rhetoric finds us in isolation amidst the mesmerizing screens of digitopia and lures us back into the noisy realm of the public agora, the marketplace of ideas. This is the realm of doxa and enthymeme, public agreements that provide the foundational material of public reasoning. Not only is rhetoric kairotic, searching by means of argument for the moment of decision. It is also doxastic, developing out of agreements about what is valuable, true, or true enough to guide action. Doxa is the starting point of rhetoric, the refinement of doxa one of its by-products. Doxa was common opinion but might better be characterized as common sense, the sense of the community—a public awareness.

Today in the Western world, we live in the comfortable assurance that rhetorical procedures such as debates, expository writing, and persuasion are goods that will persist. But these forms and forums have to be nurtured if they are to flourish. We must attend to the structures of the public sphere in which messages develop, refining contexts so as to encourage good public discourse. The rhetorician's awareness of structure, of arrangement, what Cicero called *dispositio*, calls attention to the architecture of discourse. Media theorist Clay Shirky has studied the architecture of online sites to determine which structures are conducive to helpful interactions in our new public sphere. He writes of a rhetorical tragedy of the commons: a "tragedy of the commons occurs when a group holds a resource, but each of the individual members has an incentive to overuse it." In online forums,

> the group as a whole has an incentive to keep the signal-to-noise ratio high and the conversation informative. . . . Individual users, though, have an incentive to maximize expression of their point of view, as well as maximizing the amount of communal attention they receive.

Shirky connects the question of attention to that of design. Participants seek the attention of other readers in online conversations, but their efforts often are counterproductive. Shirky wonders whether new structures for participating in an online conversation might help solve the problem of counterproductive attention seeking. The issue of attention—how it is achieved, structured, and managed—is one that a new generation of rhetorically trained designers may help to address.

Ancient rhetorical training exhibited a pronounced concern for preparing citizens to exercise phronesis in making effective practical judgments. The Athenian rhetorician Isocrates and the Roman rhetorician Quintilian were not satisfied to produce clever and entertaining speakers. Rather they wanted to develop the political cultures of Athens and Rome, respectively, by producing responsible orators who advanced the state's values. Their interest in rhetoric revolved around what Poulakos calls making "a positive contribution to the life of the audience" ("Rhetoric" 76). In *Antidosis* (c. 353 B.C.E.) Isocrates argued that

there has been implanted in us the power to persuade each other . . .
we have come together and founded cities and made laws and invented
arts; and, generally speaking, there is no institution devised by [peo-
ple] which the power of speech has not helped us to establish. (sec. 254)

This value for creating a cooperative public out of isolated individuals ought
to mark contemporary rhetorical education as well.

Cicero's canon of memory, or *memoria*, comes into play here, and in an
unexpected way. Naomi Baron is concerned that the fast pace and conversational tone of digital writing might shift our sense of "what is ephemeral and
what is durable" (166). Baron connects eloquence to memory and asks us to
consider what is *worth* remembering. Journalist Clive Thompson comments
on Baron's research: "For writer and reader alike, elegant, formal prose is
both a catalyst and a medium for certain types of thinking. If the Internet
encourages us to write primarily in 'informal spoken language,' good writing
may become rare." Thompson adds, "dull writing is neither memorable nor
engaging" ("Floating Signifiers").

Baron's research reminds us that memory is reinforced by eloquence. Thus
our public knowledge will be impoverished, perhaps even put at risk of being lost, by inelegant, monotonous language. Constructing durable public
knowledge may require renewed attention to eloquence as an aid to memory
and thus to judgment (phronesis). Flat and forgettable style does not provoke
memory and its complementary activity, careful reflection. I have two versions of Lincoln's second inaugural address in my files—one as he originally
delivered it and another I have stripped of all eloquence: where Lincoln said,
"Each looked for an easier triumph, and a result less fundamental and as-
tounding," my eloquence-depleted version reads, "Both sides thought they
would win easily, and no one thought we'd have this kind of catastrophe."
I ask the students if it matters that Lincoln's version is more eloquent. They
usually say that my eloquence-free version is easier to understand but that his
version would be remembered longer.

Conclusion

"If we can discern the large outlines of persistent forces, we can better educate
our children in the appropriate skills and literacies needed for thriving in that
world"; this quotation from *Wired* magazine founder Kevin Kelly appears in
his book *What Technology Wants* (185). Kelly is talking about educating a rising
generation for the grand forces of the powerful technologies that shape our
lives. This vision of education construed as preparation for human flourishing

is richer and nobler than fashioning competent citizens of a digital age. The method—discerning the larger outlines of persistent forces and then educating in response to those forces—I find exciting and energizing. Cultural and historical discernment is the necessary precondition of curricular development that has as its goal the human flourishing of students.

Technology affects how we live, how we think, and, as importantly, how we think of ourselves. In *You Are Not a Gadget*, Jaron Lanier makes an impassioned plea for holding on to our humanity in the face of massive technological change and dominant metaphors such as the brain as computer and the body as machine (Lanier). Rhetoric is underappreciated as a *humanizing* force in education and public life. This ancient discipline, with human concerns such as phronesis, doxa, kairos, and enthymeme at its core, recognizes individual human beings as reasoners inhabiting communities of reasoners. With its defining emphasis on audience values (doxa), practical judgment (phronesis), and arguments constructed jointly by speakers and audiences (enthymeme), rhetoric honors communities. Rhetoric also reminds us that we are shaped by language and that language enchants our lives, attuning us with kairos—the centrality of timing, the necessity of nuance, and the benefits of decorum in our lives together. In the face of technological forces imposing abbreviation, isolation, and distraction as a matter of design, we need to retrieve in our contemporary paideia the qualities on which rhetoric was founded.

Classical rhetoric has much to recommend it as a study in this age of rapid and unpredictable social, commercial, and political change. Rhetoric is the art of public *thinking*, and its appreciation for rational flexibility, adaptability, interpretation, inventiveness, expression, memory, and decorum make it essential to education in the digital age. And, in an age in which machines are occupying larger and larger swaths of the job market, it might not be such a bad idea to head in the direction that machines find it hard to go—toward narrative, eloquence, analogy, creativity, flexibility, and imaginative insight. This is what the liberal arts were about in the first place, equipping free and educated minds to achieve and express new understandings of the *human* world, while building on established cultural values. We need to reclaim that awareness today with a robust endorsement of that one, foundational liberal study that has remained contemporary in every age.

Notes

1. I employ the phrase *digital native* in a casual, even playful, sense. The label has generated a bit of controversy since its introduction around 2000. Nevertheless, the idiom seems a handy way to express an uneasiness many of us have felt in the classroom

(and elsewhere)—that our students are simply more comfortable with digital technologies than are we their teachers. Of course, embedded in my argument is my own ironic interrogation of the label: if the ancient rhetoricians can teach contemporary students a thing or two about composing and interpreting texts, as I argue they can, then maybe the Attic *logographers* and our own budding rhetoricians have more in common than "digital natives" might suggest.

2. On rhetoric in Greece and Rome, see Kennedy, *Classical Rhetoric and Its Christian and Secular Tradition*; Poulakos, *Sophistical Rhetoric in Classical Greece*; Vickers; Welch.

3. See de Romilly, *The Great Sophists in Periclean Athens*; Enos; Guthrie; Jarratt; Kitto; Ober; Poulakos, "Toward a Sophistic Definition of Rhetoric"; Rankin; Schiappa; Sutton; Walker.

4. On Protagoras's contribution to rhetoric, see Billig, chapter 3 (61–80).

Works Cited

Baron, Naomi S. *Always On: Languages in an Online and Mobile World*. Oxford UP, 2008.

Billig, Michael. *Arguing and Thinking: A Rhetorical Approach to Social Psychology*. 1987. Cambridge UP, 1996.

Carr, Nicholas. "Is Google Making Us Stupid?" *The Atlantic*, 1 July 2008, www.theatlantic.com/magazine/archive/2008/07/is-google-making-us-stupid/306868/.

———. *The Shallows: What the Internet Is Doing to Our Brains*. Norton, 2011.

de Garis, Hugo. "Globa: Accelerating Technologies Will Create a Global State by 2050." *Kurzweil Accelerating Intelligence*, 19 Jan. 2011, www.kurzweilai.net/globa-global-state-by-2050.

de Romilly, Jacqueline. *The Great Sophists in Periclean Athens*. Translated by Janet Lloyd, Clarendon Press, 1992.

———. *Magic and Rhetoric in Ancient Greece*. Harvard UP, 1975.

Enos, Richard Leo. *Greek Rhetoric before Aristotle*. Waveland, 1993.

Gagarin, Michael. *Antiphon the Athenian: Oratory, Law and Justice in the Age of the Sophists*. U of Texas P, 2002.

Guthrie, W. K. C. *The Sophists*. Cambridge UP, 1971.

Isocrates. *Antidosis*. Translated by George Norlin, 1980. *Perseus Digital Library*, Tufts University, www.perseus.tufts.edu/hopper/text?doc=Perseus%3Atext%3A1999.01.0144%3Aspeech%3D15%3Asection%3D254.

———. *Panathenaicus*. Edited by George Norlin, 1980. *Perseus Digital Library*, Tufts University, www.perseus.tufts.edu/hopper/text?doc=Isoc.%2012.30&lang=original.

Jarratt, Susan C. *Rereading the Sophists: Classical Rhetoric Refigured*. Southern Illinois UP, 1991.

Jarratt, Susan, and Rory Ong. "Aspasia: Rhetoric, Gender, and Colonial Ideology." *Reclaiming Rhetorica: Women in the Rhetorical Tradition*, edited by Andrea Lunsford, U of Pittsburgh P, 1995, pp. 9–24.

Kelly, Kevin. *What Technology Wants*. Viking, 2010.

Kennedy, George. *Classical Rhetoric and Its Christian and Secular Tradition*. 2nd ed., U of North Carolina P, 1999.

Kitto, H. D. F. *The Greeks.* 1951. Penguin, 1968.

Lanham, Richard. *The Electronic Word: Democracy, Technology, and the Arts.* U of Chicago P, 1993.

———. "Rhetorical Paideia: The Curriculum as a Work of Art." *College English*, vol. 48, no. 2, 1986, pp. 132–41.

Lanier, Jaron. *You Are Not a Gadget: A Manifesto.* Vintage, 2011.

Midgley, Mary. *Evolution as a Religion: Strange Hopes and Stranger Fears.* 1985. Routledge, 2002.

Ober, Josiah. *Mass and Elite in Democratic Athens: Rhetoric, Ideology, and the Power of the People.* Princeton UP, 1989.

Poulakos, John. "Rhetoric and Civic Education: From the Sophists to Isocrates." *Isocrates and Civic Education*, edited by Takis Poulakos and David DePew, U of Texas P, 2004, pp. 69–83.

———. *Sophistical Rhetoric in Classical Greece.* U of South Carolina P, 1994.

———. "Terms for Sophistical Rhetoric." *Rethinking the History of Rhetoric: Multidisciplinary Essays on the History of Rhetoric*, edited by Takis Poulakos, Westview, 1993, pp. 53–74.

———. "Toward a Sophistic Definition of Rhetoric." *Philosophy and Rhetoric*, vol. 16, 1983, pp. 35–48.

Rankin, H. D. *Sophists, Socratics and Cynics.* Croom Helm, 1983.

Schiappa, Edward. *Protagoras and Logos.* U of South Carolina P, 1991.

Shirky, Clay. "Group as User: Flaming and the Design of Social Software." Networks, Economics, and Culture email list, 5 Nov. 2004, art.yale.edu/file_columns /0000/2307/shirky.pdf. Archived copy.

Stanford Study of Writing, ssw.stanford.edu/.

Stone, Linda. "Continuous Partial Attention." lindastone.net/qa/continuous-partial -attention/. Blog post undated.

"Study Shows Time Spent Online Important for Teen Development." *MacArthur Foundation*, 20 Nov. 2008, www.macfound.org/press/publications/study -shows-time-spent-online-important-for-teen-development/.

Sutton, Jane. "The Marginalization of Sophistical Rhetoric and the Loss of History." *Rethinking the History of Rhetoric*, edited by Takis Poulakos, Westview, 1993 pp. 75–90.

Thompson, Clive. "Clive Thompson on the New Literacy." *Wired*, 24 Aug. 2009, www.wired.com/2009/08/st-thompson-7/.

———. "Floating Signifiers: The Internet Hasn't Killed the English Language—Yet." *Bookforum*, Jan./Feb. 2010, www.bookforum.com/inprint/016_05/5004.

Tompkins, Jane P. "The Reader in History: The Changing Shape of Literary Response." *Reader Response Criticism*, edited by Tompkins, Johns Hopkins UP, 1980, pp. 201–32.

Vickers, Brian. *In Defense of Rhetoric.* Clarendon Press, 1988.

Walker, Jeffrey. *Rhetoric and Poetics in Antiquity.* Oxford UP, 2000.

Welch, Kathleen E. *The Contemporary Reception of Classical Rhetoric: Appropriations of Ancient Discourse.* Lawrence Erlbaum, 1990.

12

Text, Design, Code: Digital Rhetoric in Academic and Professional Writing

Douglas Eyman

I will be presenting a brief overview of digital rhetoric as one of the key future directions in academic and professional writing, but I should note that many of the central concerns of digital rhetoric have been treated in depth in other contributions to this collection: James Herrick's application of classical rhetoric to students' digital composing practices, Anne Wysocki's focus on the aesthetic as rhetorical in the context of the digital, and both Wysocki's and Bazerman's focus on embodiment and the relationship between the human and the technological. In this chapter I address a contemporary view of writing that is inherently visual and digital, enacted through any number of semiotic resources that constitute a text, and composed via rhetorical methods.

I'd like to begin my consideration of future directions in academic and professional writing by positing four principles or premises that inform my view of the current state of writing studies writ large and thus serve as the starting point for thinking about what comes next. The four premises about contemporary writing practices are as follows:

- writing is visual;
- writing is digital;
- our definition of "text" is no longer solely alphabetic but includes multimedia;
- the theories and practices of rhetoric serve as the foundations for teaching academic and professional writing.

Writing Is Visual

Historically speaking, writing began with the line as a representation of a numerical value rather than as a representation of speech (Fischer; Ingold), and there have been several gestures toward thinking of writing as an extension of thought that does not require speech, from Derrida's rejection of Saussure's argument for the primacy of speech to more recent work in linguistics (Coulmas) and anthropology (Fischer). Indeed, the earliest forms of writing were designed to represent clay tokens used in accounting systems: these first writing systems used visual elements—size and shape—to take the place of the object (Schmandt-Besserat). These early forms of writing were not representations of the words that named the tokens but were visual depictions of the tokens themselves. But even in cases where text appears to be a representation of speech, the text itself is visual in nature and necessarily inhabits a visual milieu.

Contemporary forms of writing—especially in text-messaging contexts—now routinely include pictographic information that is neither text nor representative of speech in the form of emoji, which can work in concert with alphabetic text or replace it altogether (as seen in the growing number of translations of literary texts, such as *Emoji Dick*, a version of *Moby Dick* rendered entirely in emoji [Benenson]).

Anne Wysocki notes, in her chapter for this collection, that the evolution of the command-line interface into the graphical user interface "helped us see that *all* print alphabetic texts engage us visually," and she points to the affordances of word-processing and desktop-publishing software that bring to writers the power to implement layout and other design features as a part of the writing process. I would add that the visual aspect of writing is a site of rhetorical power, from the selection of typeface to the choices we make about headings, white space, and line and paragraph length.

But Wysocki's interest in the interface also helps to demonstrate the importance of the context of writing, which is itself almost invariably visual—whether black text on crisp white paper, bound within a book with a dusk jacket or cover designed to entice the reader with a visual depiction of the written content, or presented within a PowerPoint presentation (with attendant design of slides) or accessed within the highly stylized interface of the web (where visual cues are taken both from the browser interface and the interaction among text, design, and code that takes place on the web page itself).

While Diana George has argued that it has become common "to argue that writing is itself a form of visual communication" (13), the visual aspects of the text in themselves are often seen as less compelling than how to incorporate and integrate them, particularly in academic writing (the main focus of George's influential article). This renewed interest in the visual in writing studies has been advanced by writing technologies that allow writers to easily compose in text and image (as Charles Bazerman notes in his contribution to this collection when discussing technologies that support higher-order skills). These technologies can foreground the visual nature of writing by providing methods for manipulating color, size, orientation, and page design. Or the writer may choose to allow the system to determine the efficacy of the visual parameters by choosing to use preset configurations (or, rather, failing to investigate and make use of the rhetorical qualities inherent in the visual elements of writing).

One of the challenges of the visual nature of writing is that it is inaccessible to those who cannot see. So while the history and future of writing is largely bound up with its visual representation, it is also important to recognize the limitations of the visual and to note that written texts can be transfigured into aural and tactile representations via screen readers and Braille, for instance. While writing is inherently visual, it is not confined only to the visual—and these changes in form are accommodated by the other inherent nature of writing: its digitality.

Writing Is Digital

In order to make this claim, I must first note that I am using *digital* in its technical guise: the digital is embodied in any system made up of discrete values as opposed to continuous values. As I have noted elsewhere, any system made up of individual elements satisfies the technical definition, and writing, with its individual letters making up words that build larger syntactic units, is a prime example. The previous example of the earliest writing systems, which were composed of discrete physical objects subsequently represented in visual form, is also a digital system. Much as writing has always been visual, so has it been digital.

Aside from the technical definition, however, the majority of writing that is being composed and delivered in both academic and workplace settings is also digital in the more common sense of the term (that is, digital as opposed to physical, rather than digital opposed to analog). Technology mediates the vast majority of contemporary writing, whether it is composed on a word processor on a desktop or laptop, a note-taking app on a tablet, or a text-messaging system on a phone. If most of the text produced today is digitally composed (whether it resides in purely electronic form or is eventually transcribed onto paper or another physical medium), then it strikes me as imperative that scholars both research and teach writing as digital practice.

But in making this call, I must return in part to the technical definition and to another etymological foundation in order to avoid conflating the digital solely with computer and electronic technologies. Digital is also derived from *digits*, or fingers, which reminds us that writing, even when digital, is a technology that requires bodies and physical action (at least when humans compose;

for writing composed by algorithms, that embodiment is once removed but still present). Claims that the digital as a basis for written communication can erase the bodies of writers (and their attributes, such as race, class, gender, and ability) have been fairly soundly rejected both in contemporary work on digital writing practices and when taking into account the historicity of writing as having always been digital.

An Expanded Definition of "Text"

Writing (as process and product) is far more malleable in digital environments, and writers can now shape and reshape the forms of their texts. As writers, Gunther Kress argues, "individuals are now seen as the remakers, transformers, of sets of representational resources—rather than as users of stable systems, in a situation where a multiplicity of representational modes are brought into textual compositions" ("English" 87). This multiplicity of modes is made possible by the digital platforms and networked writing tools that most writers now use to compose.

In rhetorical studies, a text can be thought of as the container for an argument or persuasive discourse—and this text need not only take the shape of traditional print. Indeed, Johndan Johnson-Eilola has argued that "any object, collection of objects, or contexts can be 'read' by tracing and retracing the slipping, contradictory network of connections, disconnections, presences, absences, and assemblages that occupy problematic spaces" (33). In writing studies, then, text no longer equates to just print but can include multimedia and even multimodal elements.

Examples of writing that moves beyond traditional print abound on the web, from the circulation of memes that make arguments through the juxtaposition of a simple image and an alphabetic inscription, to visual blogging platforms like Tumblr, where the arrangement and order of images can constitute a text and form an argument. When we move beyond the notion of text as a particular form of encoding, we also make available the power of digital transformation as a means to compose and shape texts across media. One of my favorite examples is the work of Jessica Hagy, who would draw a simple diagram with a caption and legend (synthesizing alphabetic text and image) on a three-by-five-inch index card and then scan each card into a blog where others could access her typically whimsical expressions (Hagy). Moving from print to digital format allowed for both circulation and remix—one of the key affordances of an expanded view of text is that it can be composed and recomposed from any number of semiotic systems, and is not limited to just one mode of representation. And Hagy's work was taken up and remixed into a narrative arc that re-presented her index card drawings as an animated short feature, with a musical score and voice-over narration, titled *Le Grand Content* (Kogler et al.). The remix is itself text as I have defined it here, and that text was written through a networked, unplanned collaboration between the original writer and the writers who remixed her work.

Of course, if nearly everything can be seen as text, what is not text? I would suggest that there is a distinction between text as an inscription technology and as performance. Performance, as action, is not text, but the recording and inscription of performance becomes a text. Actions that are recorded become texts, but those actions that are not recorded do not become texts. Rhetorical action can take place in either or both performance and text.

Rhetorical Foundations

As a rhetorician, I tend to think that rhetoric is the foundation of all the liberal arts and sciences, but I will resist such a broad claim and instead focus on the use of rhetoric in composition and professional writing pedagogies. Rather than make the case in full here, I direct the reader's attention to Jeanne Fahnestock's contribution to this collection, which provides an argument for the value of rhetoric, aiming to "illustrate the usefulness of the rhetorical principles" in both composition and professional writing contexts.

In terms of contemporary practice, it is easy to see the role of rhetoric in professional writing—it is taught as both theory and method for the production of effective audience- (or user-)centered texts. Explicit instruction in rhetoric is less universal in the teaching of composition, but the core principles of stating one's purpose, engaging an audience, and selecting an appropriate style and medium for that purpose and audience have become central to composition pedagogy.

In light of the preceding three premises (writing as visual, writing as digital, and text as encompassing a broad range of semiotic resources placed in the service of writing), perhaps the most salient definition of rhetoric is that proposed by Robert Davis and Mark Shadle in *Teaching Multiwriting*:

> Rhetoric is a syncretic and generative practice that creates new knowledge by posing questions differently and uncovering connections that have gone unseen. Its creativity does not exclude or bracket history but often comes from recasting traditional forms and commonplaces in new contexts and questions. (103)

The methods of rhetoric are equally useful for analyzing texts and performances as for composing them; the flexibility provided by such "syncretic and generative practice" means that rhetoric aligns particularly well with new approaches to producing and understanding digital texts (many of which are composed through juxtaposition and remix, as with the example of Hagy's work). Rhetoric, however, is also a powerful tool—one that requires an attention not just to making or to understanding but also to the ethical implications of analyzing and composing rhetorically. As Lloyd Bitzer argues, "rhetoric is a mode of altering reality, not by the direct application of energy to objects, but by the creation of discourse which changes reality through the mediation of thought and action" (4). As writers, we are all in the business of altering reality, but it seems incumbent upon those of us who teach rhetoric and writing to include an ethic of care and consideration when undertaking such a potentially dangerous task.

Digital Rhetoric: Text, Design, Code

Taking into account the visual and digital nature of writing, an expanded view of what may constitute text (the product of the activity of writing), and the function of rhetorical theory and method in the teaching of writing, I believe that one of the key future directions in research, technology, and best practices in academic and professional writing is the rising field of digital rhetoric.

Digital rhetoric begins with the premise that rhetorical theories and methods can be applied as readily to technology-mediated texts as they have been to speech and traditional print forms of communication. Elsewhere I have defined digital rhetoric "as the application of rhetorical theory (as analytic method or heuristic for production) to digital texts and performances" (*Digital Rhetoric* 44), but there is quite a lot of room for delineating how these theories and methods are applied and how they might be mediated or changed through interaction with the advance of new communication and knowledge-making technologies.

In his chapter for this collection, "From *Topoi* to Tweets," James Herrick presents one of the two main approaches of digital rhetoric: the application of classical rhetoric to digital communications. Herrick argues that the practices of classical rhetoric can help students engage with argumentation and critical thinking in the environments of new media in which they live, play, and perform their identities. A number of scholars, from Richard Lanham to Jim Zappen to Barbara Warnick, have applied the terms and approaches of classical rhetoric to online writing, and a special issue of *Kairos* in 2007 focused on how Greek and Roman classical rhetoric might function for digital communications. Applying the classical appeals (ethos, pathos, logos) to new media forms as an analytic framework is often as straightforward as applying them to traditional texts, although these approaches sometimes help us to retheorize both the communication practice and the way we understand the appeals themselves in these new contexts (see, for example, Sierra and Eyman on ethos in massive multiplayer online role-playing games). And mapping the traditional canon onto new technologies is also a common starting point for such applications (table 12.1).

In that special issue of *Kairos*, however, we also published an example of an alternative approach—one that considers whether we need not merely apply but actually remediate or change classical rhetorical approaches to account for technological change that the original theorists could never have imagined. In "Re-situating and Re-mediating the Canons: A Cultural-Historical Remapping

Table 12.1. Digitizing Classical Rhetoric

Canon	*Classical Definition/Use*	*Digital Practice*
Invention	Finding available means of persuasion	Searching and negotiating networks of information; using multimodal and multimedia tools
Arrangement	Formalized organization	Manipulating digital media as well as selecting ready-made works and reconstituting them into new works; remixing
Style	Ornamentation/ appropriate form	Understanding elements of design (color, motion, interactivity, font choice, appropriate use of multimedia, et cetera)
Delivery	Oral presentation	Understanding and using systems of distribution (including the technical frameworks that support varying protocols and networks)
Memory	Memorization of speech	Information literacy—knowing how to store, retrieve, and manipulate information (personal or project-based; blogs or databases)

Source: reprinted from D. Eyman, *Digital Rhetoric: Theory, Method, Practice*, p. 65. Ann Arbor: University of Michigan Press, 2015. This work is licensed under a Creative Commons Attribution-NonCommercial-NoDerivatives 3.0 License.

of Rhetorical Activity," Paul Prior and coauthors argue that "the classical canons have always represented only a partial map of rhetorical activity . . . and that a remapping of rhetorical activity grounded in cultural-historical activity theory (CHAT) better addresses the diversely mediated practices of rhetoric."

Another key example of reframing the classical and even of pushing for new rhetorical approaches to digital communications is Collin Brooke's *Lingua Fracta: Toward a Rhetoric of New Media*. Brooke argues that the "canons can help us understand new media, which add to our understanding of the canons as they have evolved with contemporary technologies. Neither rhetoric nor technology is left unchanged in their encounter" (201).

Much of the push for rethinking how classical approaches might work for digital texts comes from the acknowledgment that texts are not just digital (and thus transportable, replicable, malleable) but are also networked. While traditional texts have always circulated within social, scholarly, and economic networks, the invention of digital networks means that texts are formed in and integrated with multiple networks, usually right from their invention. Tiziana Terranova suggests that networks themselves can lead to new forms of invention: "If the network is a type of 'spatial diagram' for the age of global communication, the self-organizing, bottom-up machines of biological computation capture the network not simply as an abstract topological formation—but as a new type of *production machine*" (100; emphasis in original).

I've come to think of writing as a kind of network(ed) *infrastructure* that undergirds (and is required for) a broad range of communication practices; that is, Terranova's "production machine" cannot engage invention without writing.

I'd like to digress for a moment and make a distinction between technical infrastructure (the physical networks, wires, computer configurations, and software protocols that make digital networks possible) and the broader sense in which I take up networks, which is more akin to the way that Ulises Mejias describes networks "as assemblages of people, technology and social norms" that "arrange subjects into structures and define the parameters for their interaction, thus actively shaping their social realities" (qtd. in Langlois et al. 429). In the case of the infrastructural elements of all networks, I also agree with Jeff Grabill's contention that "infrastructure . . . is not stable, fixed—visible even—but rather emerges—becomes visible and meaningful—through use" (15).

While I think that it is actually critical for writing studies scholars to account for the affordances and constraints of technical infrastructure (for an extension of this argument, see DeVoss, Cushman, and Grabill in "Infrastructure and Composing"), my interest here is linking writing itself as necessary for the formation of both social and technical networks.

In other venues Cheryl Ball and I have argued that the work of writing studies researchers requires particular forms of infrastructure—scholarly, social, and technical—in order to become contributions to our field (Eyman and Ball, "Composing" and "Digital Humanities Scholarship"); these infrastructures are networks of practice that need to be in place in order to circulate our own writing. I would add that writing itself (and digital writing in particular) requires specific infrastructural elements to be most effective, and I break down these elements into categories of text, design, and code.

Text

As noted above, *text* has now taken on a fairly broad meaning and often includes multiple media elements working together. This is not a new phenomenon, as many print texts include design elements and images working in concert with words. But we now have a much broader array of semiotic systems to choose from in encoding our thoughts and ideas.

In terms of infrastructure, it is useful to explicate what text does—the work that it accomplishes as a text. To this end, I rely on Robert de Beaugrande and Wolfgang Dressler's approach to text as a "communicative event" (1). While de Beaugrande and Dressler catalog a number of features of text that could help to articulate its infrastructural application, I favor Ali Darwish's version, which argues that text comprises six layers: textual, contextual, cultural, temporal, intentional, and intertextual (155–56), because these elements work well with the notion of text as networked components of the larger activity of writing.

We can also see text as the inscription of performance (the temporal aspect that Darwish provides in his schema); thus, text captures the performance of writing even as it instantiates it in permanent form. This view of text as an infrastructural component of the activity of writing refocuses our attention on just how we might define and understand text. (In my reviews of writing studies research, the term is often taken as self-explanatory and not interrogated in any way). I would suggest that a renewed attention to just what we mean by *text*—its affordances, boundaries, capacities, and functions—is one research strand that can be taken up by digital rhetoric.

Design

Design becomes a critical faculty for the production of digital texts, particularly when our notion of text now encompasses image, animation, video, and audio, as well as print. But even in writing that is only print, design elements can play an important role in support of the rhetorical power of the text. Design elements such as typeface and layout should be taught alongside grammar, style, and organization as key rhetorical methods for helping match the message to the intended audience.

Kathryn Northcut and Eva Brumberger argue that most approaches to teaching composition often fail to see these design elements as rhetorical, relying instead on the "crystal goblet" approach:

> The metaphor of the crystal goblet compares the typographic features of text to a fine crystal goblet: invisible, or decorative, if done well, intended solely as a vessel for the contents, which remain the same regardless of the vessel. . . . Such a belief precludes the need to treat typography as rhetorical. (462)

As with text, it is my hope that digital writing and digital rhetoric can bring attention to design as an integral component of writing (and not just new media writing but all digitally mediated writing).

As we move toward more visually oriented writing practices (both in professional contexts and in academic writing), design begins to play a much stronger role than it has traditionally been granted in the teaching of writing. One way into design as a rhetorical component of writing, then, is to look at document design in the composition class.

In my work as editor of *Kairos*, I have championed the idea of text-as-*designed* discourse. In *Literacy in the New Media Age*, Gunther Kress proposes a theory of text that includes three categories of text (aesthetically valued, culturally significant, and mundane), each of which is expressly the result of specific design choices:

> text is based, however imperfectly, on the understandings of design: an understanding of what the social and cultural environment is into which my text is to fit, the purposes it is to achieve, the resources of all kinds that I have to implement and realise my design, and the awareness of the characteristics of the sites of appearance of that text. (120)

These design choices are critical for new forms of digital scholarship (such as those we publish in *Kairos*), but they are just as powerful (when made explicit) for all forms of writing.

Code

Code is the underlying structure that has to function properly in order for a digital text to achieve its design goals and support its rhetorical functions. In the field of digital rhetoric, there is currently a great deal of work that focuses on the role of code, much of which takes as its starting point Ian Bogost's notion of procedural rhetoric, "the art of persuasion through rule-based representations and interactions, rather than the spoken word, writing, images, or moving pictures" (ix). Digital rhetoric scholars are interested in how code (and algorithms) interact with text and design to perform rhetorical functions, and there are intersections between this work in digital rhetoric and critical code studies.

In a recent article, Cheryl Ball and I have argued that

> [c]ode is, in a way, analogous to grammar—in order to function properly it needs to adhere to certain standards: it must be well-formed and conform to a formal register that is (generally) enforced by the systems that interpret and execute the code. ("Composing" 116)

While code is an infrastructural element of digital writing, it is also a necessity for the development and implementation of technical networks as well. The challenge for digital rhetoric, as I see it, is not to lose sight of the function of code in the domain of writing.

The Future Is Now

By the time this collection of essays is published, we will already be working to understand new challenges to writing studies presented by changes in technology. We will likely see a continued shift from larger to smaller interfaces for everything from websites to mass media, as mobile and portable devices continue to outpace and out-innovate televisions, desktop computers, and laptops. It seems likely that more systems will use "natural user interfaces" such as gestures that are read by our devices' sensors (also an opportunity to bring classical rhetoric approaches to gesture and oratory into digital composing practices). And we'll be living in a literal *internet of things* as more objects in our homes and workplaces become part of the (technical) networks of the internet.

Even with these changes—or perhaps spurred on by them—I believe that a focus on writing (as I have configured it here) as both a necessary element of and built on infrastructural networks will help us to see the value and centrality of writing, not as an outdated skill that will be supplanted by new media forms and artificially intelligent author-robots, but as a foundational activity that drives and makes possible these new digital communication practices. That means we must address writing in both academic and professional contexts in terms of its visual and digital nature and we must encourage and train our students to use rhetorical methods to compose texts across a range of media, modes, and networks.

Composition and professional writing pedagogies can both take up the infrastructures of the writing approach I've outlined here, addressing the role of text, design, and code as composed through the heuristics of rhetorical theory. In so doing, we can also reclaim context as a central consideration and help students to understand better the interplay among interface, network, and their own writing practices.

Even as we continue to theorize and better understand digital rhetoric (for both production and analysis), I believe it is also important to historicize our notion of what writing is and what it does even as we create new ways of making meaning through digital writing practices. And we must also not lose sight of the complications of writing in a digital networked world: What is the place of the body in digital rhetoric? How do we accommodate readers/users who have different capabilities from those in the mainstream to ensure that new forms of writing remain as widely accessible as possible? And what kind of ethical training should we attend to when teaching students to alter reality through the powerful rhetorical forms of discourse we hope they will employ?

If we attend to these questions as we continually refine instruction in the fundamentals of digital rhetoric practice, I believe that writing and writing instruction will continue to have an important role to play in both making and negotiating all of our possible futures.

Works Cited

Benenson, Fred, editor. *Emoji Dick.* 2010. Self-published.

Bitzer, Lloyd. "The Rhetorical Situation." *Philosophy and Rhetoric*, vol. 1, 1968, pp. 1–14.

Bogost, Ian. *Persuasive Games.* MIT P, 2007.

Brooke, Collin Gifford. *Lingua Fracta: Toward a Rhetoric of New Media.* Hampton Press, 2009.

Coulmas, Florian. *Writing Systems: An Introduction to Their Linguistic Analysis.* Cambridge UP, 2003.

Darwish, Ali. *Optimality in Translation.* Writescope, 2008.

Davis, Robert L., and Mark F. Shadle. *Teaching Multiwriting: Researching and Composing with Multiple Genres, Media, Disciplines, and Cultures.* Southern Illinois UP, 2007.

de Beaugrande, Robert, and Wolfgang Dressler. *Introduction to Text Linguistics.* Longman, 1981.

Derrida, Jacques. *Of Grammatology.* Johns Hopkins UP, 1974.

DeVoss, Dànielle Nicole, et al. "Infrastructure and Composing: The *When* of New-Media Writing." *College Composition and Communication*, vol. 57, no. 1, 2005, pp. 14–44.

Eyman, Douglas. *Digital Rhetoric: Theory, Method, Practice.* U of Michigan P, 2015.

Eyman, Douglas, and Cheryl Ball. "Composing for Digital Publication: Rhetoric, Design, Code." *Composition Studies*, vol. 42, no. 1, 2014, pp. 114–17.

———. "Digital Humanities Scholarship and Electronic Publication." *Rhetoric and the Digital Humanities*, edited by Jim Ridolfo and William Hart-Davidson, U of Chicago P, 2014, pp. 65–79.

Fischer, Steven Roger. *History of Writing*. Reaktion Books, 2004.

George, Diana. "From Analysis to Design: Visual Communication in the Teaching of Writing." *College Composition and Communication*, vol. 54, no. 1, 2002, pp. 11–39.

Grabill, Jeff. "Infrastructure Outreach and the Engaged Writing Program." *Going Public: What Writing Programs Learn from Engagement*, edited by Shirley K. Rose and Irwin Weiser, Utah State UP, 2010, pp. 15–28.

Hagy, Jessica. *Indexed*. indexed.blogspot.com.

Ingold, Tim. *Lines: A Brief History*. Routledge, 2007.

Johnson-Eilola, Johndan. "Among Texts." *Rhetorics and Technologies: New Directions in Writing and Communication*, edited by Stuart Selber, U of South Carolina P, 2010, pp. 33–55.

Kogler, Clemens, et al. *Le Grand Content. YouTube*, 13 Jan. 2007, www.youtube.com /watch?v=1WWKBY7gx_o.

Kress, Gunther. "English at the Crossroads: Rethinking Curricula of Communication in the Context of the Turn to the Visual." *Passions, Pedagogies, and 21st Century Technologies*, edited by Gail Hawisher and Cynthia Selfe, Utah State UP, 1999, pp. 66–88.

———. *Literacy in the New Media Age*. London: Routledge, 2003.

Langlois, Ganaele, et al. "Networked Publics: The Double Articulation of Code and Politics on Facebook." *Canadian Journal of Communication*, vol. 34, no. 3, 2009, pp. 415–34.

Lanham, Richard. *The Electronic Word: Democracy, Technology, and the Arts*. U of Chicago P, 1993.

Northcut, Kathryn M., and Eva R. Brumberger. "Resisting the Lure of Technology-Driven Design: Pedagogical Approaches to Visual Communication." *Journal of Technical Writing and Communication*, vol. 40, no. 4, 2010, pp. 459–71.

Prior, Paul, et al. "Re-situating and Re-mediating the Canons: A Cultural-Historical Remapping of Rhetorical Activity." *Kairos: Rhetoric, Technology, Pedagogy*, vol. 11, no. 3, http://kairos.technorhetoric.net/11.3/binder.html?topoi /prior-et-al/mapping/index.html.

Schmandt-Besserat, Denise. "The Envelopes That Bear the First Writing." *Technology and Culture*, vol. 21, no. 3, 1980, pp. 357–85.

Sierra, Wendi, and Doug Eyman. "'I Rolled the Dice with Trade Chat and This Is What I Got': Demonstrating Context-Dependent Credibility in Virtual Worlds." *Online Credibility and Digital Ethos: Evaluating Computer-Mediated Communication*, edited by Moe Folk and Shawn Apostel, IGI Global, 2013, pp. 332–52.

Terranova, Tiziana. *Network Culture: Politics for the Information Age*. Pluto Press, 2004.

Warnick, Barbara. *Rhetoric Online: Persuasion and Politics on the World Wide Web*. Peter Lang, 2007.

Zappen, James P. "Digital Rhetoric: Toward an Integrated Theory." *Technical Communication Quarterly*, vol. 14, no. 3, 2005, pp. 319–25.

13

A Beauty for Informing Digital Bodies

Anne Frances Wysocki

This writing considers how digitality has opened up potentials for us not only to work with visuals but to see how we see and how we sense more generally—all of which concerns have ethical implications for how we sensuously compose each other and our worlds in writing. This chapter, therefore, considers broad questions underlying current approaches to teaching writing and—while offering a specific heuristic for teaching writing for and with our senses—addresses the future of writing studies broadly.

For several decades now we've been worrying the old dichotomies, tallying the damage from seeing only *male* or *female*, *reason* or *emotion*, *white* or *black*, *mind* or *body*, *culture* or *nature*, *discourse* or *stuff*. What gets categorized on either side of *or* in these dichotomies often ends up being seen through some combination of the other labels on the same side of the divide; we've noted too that this sorting happens neither gently nor neutrally: judgments, often harsh, always evaluative, accrue to any action or anyone seen through those limited (and limiting) possibilities. And so we have been working to bring bodies and stuff back to more positive attention, to try to talk about them otherwise, when the talk has for many years been on the first side, on the side of the abstract and rational.

We see such work around bodies, materiality, and texts in, for example, a recent analysis of the movie *Lost in Translation*; in that analysis Brian L. Ott and Diane Marie Keeling seek to understand the movie's "sensual and affective experiences fostered by its aesthetic dimensions," while noting that they do "not quote (or analyze the meaning of) a single line of dialogue" in the movie (377). Ott and Keeling argue that rhetorical bridges need to be built "between interpretive knowing and experiential knowing": they "urge rhetorical critics interested in materiality to continue developing critical practices that attend to how sensory systems such as sight, hearing, smell, taste, and touch operate experientially" (379). Similarly, in another recent article, Gregory Clark uses a Burkean analysis of Hawaiian song to argue for "a usable concept of a rhetorical aesthetic"; relying on a distinction between rhetoric and aesthetics that he claims is too simple but nonetheless useful, Clark argues that "if rhetoric involves primarily something like thought and aesthetic involves primarily something like feeling, then we probably need to employ both as we try to make sense of the identities we derive from our ways of life in space and place" (267). For Clark, it is through aesthetic experiences that "people compose identity and sustain community from the resources of immediate and symbolic experience" (269).

My essay—responding to a request to speak "on the current and future impact of visual and digital writing on academic and professional writing"— parallels the concerns of the articles above and of similar work concerned with the overlaps of aesthetics, material rhetorics, digital materiality, and the teaching of academic and professional writing. (See, for example, Hawhee, Holloway-Attaway, Knight, or Lindhé.) My essay considers how digitality has helped us see visuality in writing and so has helped us feel other senses at work in all composing. This essay consider how sensing is always evaluative, always ethical—and so this essay comes at the materiality of rhetorical practices by considering aesthetics and how we work among our senses, our bodies, our worlds, and our ways of communicating. This essay comes to the aesthetic, therefore, not because aesthetics is about what is pretty or because aesthetics can help us produce more persuasive digital texts (although aesthetics can certainly do all that) but because aesthetic practices have real and potentially painful ethical consequences, especially when we ignore the aesthetic. This essay offers a modest heuristic for using our senses critically and rhetorically in teaching about, and with, writing.

* * *

I start, then, with how digitality provides perspective for glancing back on how we once acted as though academic writing sought embodiment without sense.

Some of you will remember the command-line interfaces and blinking block letters of the earliest consumer computers—and so will remember how computers only took off as consumer devices when command-line interfaces were replaced by graphical interfaces featuring WYSIWYG (what you see is what you get) like those of the first Macintosh computers. Such interfaces have helped us see that *all* print alphabetic texts engage us visually; read Lester Faigley's words from 1999, for example: "literacy has *always* been a material, multimedia construct, even though we are only now becoming aware of this multidimensionality and materiality because computer technologies have made it possible for many people to produce and publish multimedia presentations" (175–76). Once we could move from the typewriter's constraints of one size of one typeface, and once we could move from software like WordStar (where formatting type meant coding) to software like InDesign or even Word, in which anyone can choose typefaces and sizes, we see the rhetoricality of typefaces. But we also see the broader rhetoricality of page layout itself, and anyone who can afford and learn the software can engage with chunking type into columns or callouts and inserting photographs or illustrations or can turn away from print completely to video or animation, with or without alphabetic engagement. But back when the technologies of print production depended on printing presses and typewriters, the knowledge of production belonged only to graphic designers with the expertise of years of training in, and so years of enculturation in, design and printing press intricacies. Now, as Faigley noted, anyone with sight, dexterity, and access to words, visual objects, and a computer can produce what once took a long apprenticeship and so can also produce free from the values and conventions implicit in those earlier apprenticeships.

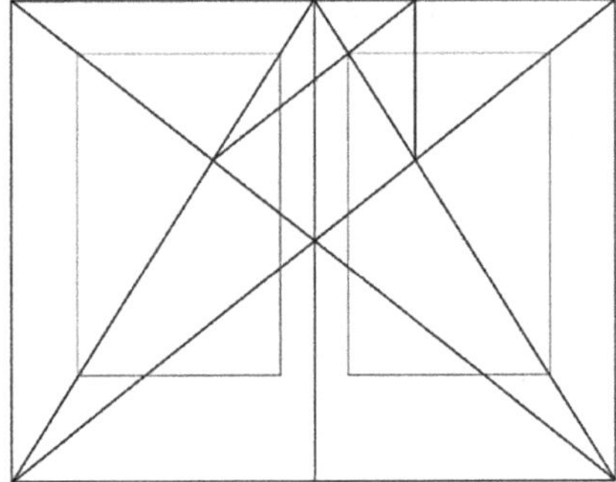

Figure 13.1. *Left,* David Carson layout, *Ray Gun* magazine, 1994. *Right,* Golden section drawn over page spread, for teaching book designers about classic page design.

Those changes in communication technologies thus also helped make visible those values and conventions, which cover so much more than typefaces. When in the 1980s young-gun designers like David Carson began experimenting with new software, graphic design journals were filled with arguments pitting classical approaches against experimentation (see Bierut et al.). Classical approaches said layout was to disappear in order to free readers' attention to focus on the ideas, for which the design was (or should be) only a neutral carrier; classical convention designed pages whose black ink was not to be seen because the page's materiality was to be transcended into the realm of ideas.

Those experimenting with the then-new possibilities of page layout software argued instead that design should reflect its time and could be personal and that design carries weight in how readers read, understand, and act. These arguments (echoing the old classical versus romantic dichotomy in art history) hinged on whether a readable page resulted from objective, apparently god-given and/ or human-physiologically-tied layout principles or, instead, from convention and from readers becoming accustomed to habitual practices with type and layout. These arguments led Richard Lanham to his 1995 distinction between "looking through" the page to ideas and "looking at" a page's surface to focus on its materiality and timeliness; in 2000 Bolter and Grusin offered their similar distinction between "immediacy" and "hypermediacy" for all media. Such arguments and distinctions helped us see the historicity and contingency of page (or any) design and the cultural functions of page (or any) design—and they helped us see *seeing*: seeing as a learned process, seeing as a process of embodying particular bodies within particularly articulated belief sets.

It's been twenty years already since Martin Jay argued in *Downcast Eyes* that we in the West have inherited a doubled notion of vision: he argues that we can understand vision either as speculation ("as the allegedly pure sight of perfect and immobile forms with 'the eye of the mind'") or as observation ("as the impure but immediately experienced sight of the actual two eyes" [29]). Jay argues that this split has been "instrumental in elevating the status of the visual in Western culture," for "when one of those alternatives was under attack the other could be raised in its place. In either case, something called vision could still be accounted the noblest of the senses" (28–29). That is, if we believe we can see through the black-ink letters on a page to a meaning that lies immaterially beyond or *through* the page, it is because we believe that seeing ought to be aligned with philosophy, with reason, with immaterial words, with pages designed to be transparent; if instead we wish to look *at* the page, then what is at work are contingent demands of our bodies, matter, emotion, and rhetoric. (For other texts on the historical and locational contingencies of seeing, see, for example, Foster, Levin, or Virilio.)

Consider the historicity of page design, then, to consider how seeing gets shaped in interaction with the values, conventions, and judgments present in any design. Johanna Drucker describes how graphic design became a professional field in the West in the years of industrialization following World War II. Drucker argues that we cannot separate our current design values of organization, efficiency, and consistency from the values of rationalization and standardization in twentieth-century industrial practices; we cannot separate the values of organization, efficiency, and consistency from pushes toward shaping a standardized workforce. These pushes are therefore inseparable from the look and lessons of the standardized pages of contemporary academic books: when we engage with such pages, we experience and so take up, without explicit instruction, the values and judgments of rationalization and standardization. Or consider how Kress and van Leeuwen argue that our attachment to what they call the "densely printed page"—the page of nothing but alphabetic text—connects to capital's entrenchment at a particular time in the development of texts for mass consumption; this process, they say,

> began in the late nineteenth-century mass press, in a context in which the ruling class, itself strongly committed to the densely printed page, attempted to maintain its hegemony by taking control of popular culture, commercializing it, and so turning the media *of* the people into the media *for* the people. Their own comparable media—"high" literature and the humanities generally—became even more firmly founded on the single semiotic of writing. Layout was not encouraged here, because it undermined the power of the densely printed page as, literally, the realization of the most literary and literate semiotic. The genres of the densely printed page, then, manifest the cultural capital ... controlled by the intellectual and artistic wing of the middle class. (185; emphases theirs)

Kress and van Leeuwen go on to argue that pages attentive to layout and variety are aimed at "'the masses,' or children" (186).

Please do not hear here a McLuhanesque we-are-determined-by-our-media argument. I do not describe a one-directional process where our media make us. Instead, notice how, in what I've described, our media shift in close feedback loops with social, cultural, economic, and political processes—and in close relations with our values and (how we understand) our bodies. In other words, we learn to see in part through what we are shown to see as we grow up and through how what we are shown requires us to use our eyes. Consider how you, as a teacher, would likely squirm to see a student hand in an academic paper hand-drawn on paper bags: imagine the hard work you would have to do to read such a paper and find quality in it without becoming distracted by your immediate and visceral judgment—based only on the look of the writing—that such a shambling presentation would have to be an intellectual as well as a visual mess.

All of that is what I meant earlier when, in considering how computer screens and software encourage us to see our pages, I said that they also encourage us to see ourselves seeing: we can see that we see different pages differently; we can see that in those different kinds of seeing there's a whole lot more than the local effects of any single page at work. (Again, I want to note here that I do not want to sound as though I am saying computers or digitality caused us to see this; these critiques of vision have developed in overlapping parallel with the work of the last fifty years in identity studies, cultural studies, and any other academic considerations of embodiment.)

And so, when we do become alert to seeing as a contingent, historical process, we can also become alert to how others might see differently, how others might be differently embodied. We can become alert to consequences of *not* accepting the contingency of our senses; we can become alert to consequences of believing and acting as though our sight, like our bodies, were natural, universal, unmediated. We can become alert to how our senses involve us, without thought, in distaste and disdain. For one example, consider Brenda Farnell's study of Plains Indian sign talk: Farnell argues that nineteenth-century (and later) Europeans saw the Plains nations' use of sign talk as an indication of their low evolutionary status. Farnell sees the disdain for sign talk (inseparably tied to the disdain for the Indians) arising out of the European dichotomizing of word and picture, for

> movement systems have been labeled primitive and robbed of validity in relation to spoken language systems because of the predominance of iconic signs [in them]. In the privileged semantico-referential category of traditional linguistics, signs are supposed to be arbitrary and symbolic, not iconic. (56)

Or consider Oliver Sacks writing about how people who are deaf in the United States were long prevented from learning sign languages because others did not perceive such nonwritten, nonalphabetic language as "real" language—with the result that many people who were deaf were prevented from learning any language. Or consider Pierre Bourdieu's arguments in *Distinction: A Social Critique of the Judgment of Taste* about the distinctive sensing cultures of different classes in France and the results thereof:

> tastes are perhaps first and foremost distastes, disgust provoked by horror or visceral intolerance ("sick-making") of the tastes of others. "De gustibus non est disputandum": not because "tous les goûts sont dans la nature," but because each taste feels itself to be natural[,] . . . which amounts to rejecting others as unnatural and therefore vicious. (56)

Bourdieu ends this passage by noting that "aesthetic intolerance can be terribly violent" (56).

Our un- or preconscious tendencies to judge skin color, gender, or age now get named *hidden bias*, *blind spots*, *implicit social attitudes*, or *unconscious bias*. On the Project Implicit website (founded in 1998 by researchers from Harvard, the University of Virginia, and the University of Washington and now with an international network of collaborators), one can take a test to check one's implicit attitudes toward black or white faces by pushing a button quickly to associate positive or negative terms with a photograph of a black or a white person,[1] and the compiled results for everyone who has taken the test indicate that "more than 70% associated 'good' with White faces more easily than with Black faces" (Nosek et al. 154; the writers also consider the validity of such tests). Other researchers have tested implicit beliefs by sending résumés to employers and university science faculty. The résumés were the same except that one had a name that sounded African American and the other one white; those tasked with hiring made hard judgments based on that name only:

> Job applicants with African-American names get far fewer callbacks for each resume they send out. Equally importantly, applicants with African-American names find it hard to overcome this hurdle in call-backs by improving their observable skills or credentials. (Bertrand and Mullainathan 1011)

Another study of a similar design tested applicants with female and male names:

> Our results revealed that both male and female faculty judged a female student to be less competent and less worthy of being hired than an identical male student, and also offered her a smaller starting salary and less career mentoring. (Moss-Racusin et al. 16477)

Such judgments result from enculturation, surely: they result from our growing up within particular families and neighborhoods; from all the conversations we've had with others; from all the television, movies, or videos we've ever watched; and from all the classrooms in which we've been taught and taught—but they also result from how, in all this talking and watching and listening and hearing, we have learned to have our particular senses that make such quick, nondiscursive, apparently natural but surely visceral judgments. Embodiment and enculturation are inseparable: becoming a sensing being is also about becoming an evaluating being.

The developing field of cultural neuroscience draws on experimental evidence suggesting that our senses are not fixed biological processes but rather are plastic and mutable, developing in the bath of culture. One review article in this field begins its section on recent research into perception by noting different cultural styles of perceiving:

> The neural substrates of human perception might seem more or less universal; after all, people in all cultures face the same basic perceptual challenges (e.g., tactile discrimination, object recognition). However, recent neuroimaging research has revealed a set of (perhaps surprising) cultural differences in the neural mechanisms subserving various perceptual domains, including object processing, colour discrimination, and taste.
>
> Behavioural studies widely suggest that East Asians and Westerners apply different "perceptual styles" to the task of decoding visual scenes. Specifically, Westerners tend to focus on objects (in an analytical, context-free manner), whereas East Asians tend to focus more on contexts, relationships, and backgrounds. (Ames and Fiske 72)

Another review article discusses how the brain adapts to culture:

> Cultural practices adapt to neural constraints, and the brain adapts to cultural practice. Other circuits are wired as a result of learning, particularly implicit learning. In this latter respect, the brain is a cultural sponge—indeed, possibly, the organ of culture. It internalizes the structural regularities of its environment within the parameters of innate and developmental constraints, and it employs these internalized representations to facilitate interaction with the physical and social world. (Ambady and Bharucha 342)

No matter where you stand on the difficulty of bridging bodies and culture using brains scans from functional MRI, there is also the evidence of cultural practices themselves. Other cultures exist that do not (for example) transcendentally privilege sight as we do in the West. Other cultures exist that ground their epistemologies and ontologies in other senses, in differing hierarchies and relationships of senses, in senses we might not even consider to be senses.[2] The Incas relied on hearing as their grounding sense, as other cultures have relied on the sense of smell or heat (Classen, *Worlds*); people can "think in sound" and "think in touch" (Howes 10) as opposed to sight, and Farnell describes how smells "talk" and movement works rhetorically in Assiniboine (122–23). (For more on cultural histories and constructions of sense, see, for example, the other works of Howes and Classen as well as the books of Haverkamp, Jütte, or Malnar and Vodvarka.) The possibilities of different

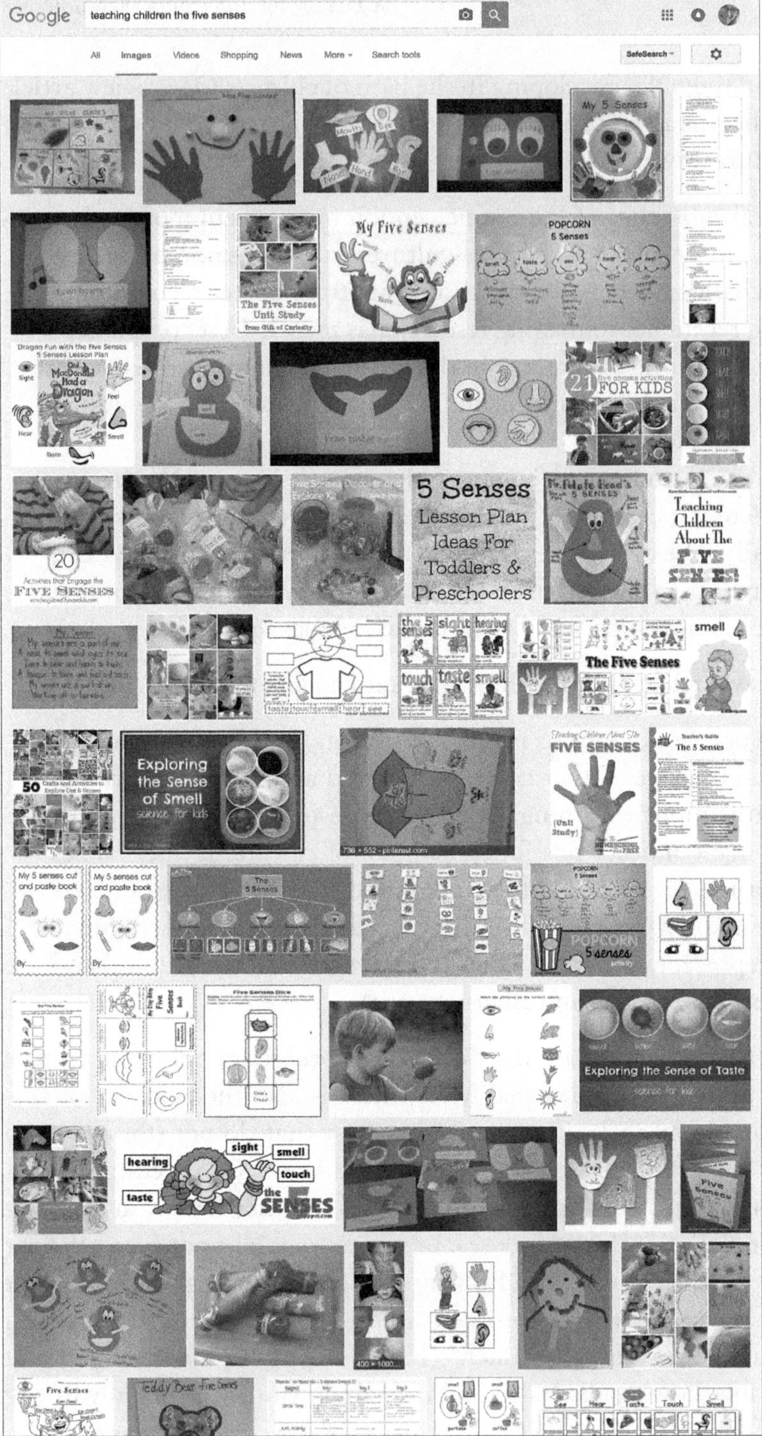

Figure 13.2. The top of my (long) page of Google Images search results for "teaching children the five senses."

sensuous engagements with the world are wide and fascinating—and can help us consider how we have come to have the bodies and related aesthetics we do in these digital times and spaces.[3]

One result of how we have shaped digital processes and tools for writing has been, then, to see and so make available for critique the kind of seeing that articulated (and still articulates) to an academic print culture that valorizes pages of nothing but rectangles of restrained typography. And so we can understand that—when we do not pay attention to the visual as a rhetorical component of our writing and we attempt then to work only with those pages we see "through"—we are actively engaged in disembodying ourselves (or, rather, we are actively engaged in acting as though disembodiment were possible).

Another result of this looking back over these values of print culture can be to learn more broadly the contingency and articulations of our sensing, as I have described above, and to learn the larger ethical implications for sensing within our particular cultural formations, as I began to articulate above. Bourdieu, for example, speaks of cultural competence, arguing that it

> results from the unintentional learning made possible by a disposition acquired through domestic or scholastic inculcation of legitimate culture. This transposable disposition, armed with a set of perceptual and evaluative schemes that are available for general application, inclines its owner toward other cultural experiences and enables him to perceive, classify, and memorize them differently. (28)

For Bourdieu, "In identifying what is worthy of being seen and the right way to see it," one is guided by the "whole social group" (28). Or, as Rancière writes,

> I call the distribution of the sensible the system of self-evident facts of sense perception that simultaneously discloses the existence of something in common and the delimitations that define the respective parts and positions within it. (12)

With the notion "distribution of the sensible," Rancière names how we each grow up in a culture that values a certain organization of our senses—an organization shaped by what we are taught our senses are and how they function—and so we grow up into systems where we sense what we are taught to sense and where we don't sense what we are not taught to sense; there is judgment and evaluation attached to each use and not-use. For Rancière, the distribution of the sensible combines the political with the aesthetic, for the distribution of the sensible is a delimitation of spaces and times, of the visible and invisible, of speech and noise, that simultaneously determines the place and the stakes of politics as a form of experience. Politics revolves around what is seen and what can be said about it, around who has the ability to see and the talent to speak. . . . (13)

If Bourdieu and Rancière are too recent for you, consider Marx's claim that the "*forming* of the five senses is a labour of the entire history of the world down to the present": for Marx, our bodily senses are what engage us socially because they engage us within and in the world, and our senses can be cultivated toward human pleasure and community—or not.

If ethics is about our communal lives (to go back to the root sense of the Greek ἦθος), then our sense formations tied to culture cannot be separated from the choices and judgments we make together grounded in our shared values. My arguments up to now depend on how media and sensory regimes entwine, and so our media, and how they engage our senses and how we respond to and work with or against those engagements, are also tightly woven into our ethics.

One implication of my claims above is that—given the tight imbrication of bodies, senses, media, and culture—changes in our media should participate in the feedback loop of the imbrication: changes in our media should articulate with changes in our sensing and so in our senses of our bodies and each other;

Figure 13.3. Engaging with nature: Pokémon Go played in Lake Park, Milwaukee, in summer 2016. Courtesy of FoxNews6.

what we choose to make with our media has effects on who we are as sensing interacting beings. I've written elsewhere about claims made in the last decade by media theorists about new media art and new sensory possibilities ("Unfitting Beauties"). Mark Hansen argues that because they can engage our senses in unexpected or new ways, digital art pieces can broaden what we might call the "sensory commons," the space we human beings share because of our particular embodiment; for Hansen, this is because digital technologies can allow us to use our senses differently than before and so expand our sensing possibilities—and can expand the range of how we interact with others and thereby expand "the agency of collective existence" (20). Similarly, Anna Munster claims that "the aesthetics of technologically inflected, augmented, and managed modes of perceptions is also about relations to others in the socius" (151); her argument points to a particular need to attend to that digital inflection, augmentation, and management of the senses. Finally, Oliver Grau argues that the "processes of digitization create new areas of perception, which will lead to noticeable transformations in everyday life" (347). These writers all argue that—with the digital—what we know about the world through our senses (not necessarily at the level of the discursive) becomes the ground for opening up the potentials of how we live together, socially, ethically. Although these writers ground their arguments in artworks, which would seem to be an obvious turn to the aesthetic, art is only a special case of aesthetic experiences. Instead, I turn to aesthetics now to consider not art but rather the quotidian visual and digital (and other) composing we do in writing classrooms. Because aesthetics, as theory, concerns our sensory engagements, and because what we do in writing classes does now so visibly and often with auditory accompaniment engage us with the senses in ways that predigital writing so avoided, turning to aesthetics here, to consider what it has to offer us as teachers of writing, matters.

* * *

From the history of aesthetic theory, and from its whispers throughout my writing so far, I make the following five points explicit, in order to build, finally, what I hope is a useful heuristic to fold into current teaching about academic and professional composing and writing:

1. Aesthetics is about sense perception and, more explicitly, about how we bridge sense perception and cognition.
2. Aesthetic judgments change over time and place.
3. Aesthetic judgments are evaluations.
4. All texts are aesthetic.

1. Aesthetics is about sense perception and, more precisely, about how we bring sense perception together with cognition.

Although much consideration that we might call aesthetic started with the ancient Greeks and has happened in non-Western traditions, the term *aesthetics* itself did not appear until mid-eighteenth-century Germany. In 1735, in *Reflections on Poetry*, Alexander Baumgarten first used the term in discussing how "things known are to be known by the superior faculty as the object of logic" while "things perceived [are to be known by the inferior faculty as the object] of the science of perception, or aesthetic" (78). Baumgarten distinguishes between reason/logic and perception so to define poetry as "a perfect sensate discourse"; he explains that a "sensate discourse will be the more perfect the more its parts favor the awakening of sensate representations" (39). Later in the eighteenth century, Kant also distinguished reason from the senses, exploring the first as "understanding," the faculty of concepts, thought, and discursivity; he explored the second as "sensibility," the faculty of intuitions, perception, and mental imagery. Unlike Baumgarten, however, for Kant the aesthetic is not equal to sensibility; instead, in *The Critique of Judgment*, Kant defines the aesthetic as the faculty for bringing together understanding and sensibility. For Kant, aesthetic judgments bring together our thoughtful, discursive capacities with what we experience sensuously. Aesthetic judgments judge—evaluate—how it feels to be in the world. They are the cognition of feeling. They are a reminder that we are bodies.

This function of aesthetic judgments continues. In 2012 Sianne Ngai defined aesthetic judgments as "evaluations based on feelings related to how things appear" and so as an "imbrication of the affective and the conceptual that underlies all judgment" (132). For Ngai, any aesthetic judgment functions not simply as "the spontaneous, feeling-based act of judgment but [as] that judgment's discursive and narrative aftermath" (170). In a very different field, in a 2013 book on "the neuroscience of aesthetic experience," G. Gabrielle Starr offers a long description of current neurophysiological understandings of aesthetic judgments; she describes how

> in general anatomical terms, neural activation moves from the sensory cortex forward through the basal ganglia (reward processes) and toward the hippocampus and amygdalae (memory and emotion—though these functions are not carried out exclusively in these structures). Activation in the orbitofrontal cortex follows, but there are interactive loops that reach between these frontal areas and the basal ganglia so that higher-order, complex processes of cognition,

and emotional and reward processes, may continually feed into one another. (24)

The aesthetic has been and continues to be an approach for studying how we plait cognition and our feelings of and about objects.

2. Aesthetic judgments change over time and place.

Charles Hartshorne, a process theologist working in the latter half of the previous century, categorized in a diagram what he claimed to be all possible aesthetic judgments; his major categories draw on what one might call "classical"—art historical—aesthetic judgments (Spuybroek). Hartshorne's paired categories are the neat and the ugly, the commonplace and the tragic, the pretty and the sublime, and the superb and the comic, all of them centered on the beautiful. Into those various categories, Hartshorne intends to have us subsume other judgments, judgments about (for example) the ridiculous, the cute, the pretty, the cool, or the elegant. In comparison, consider the judgments of disgust or "feeling sick" that Bourdieu describes the differing classes making about each other's tastes, or consider how Ngai argues for the time-based specificity of judgments when she writes that

> contemporary zaniness, cuteness, and interestingness are, at the deepest level, about performance, commodities, and information. ... [B]y calling forth specific powers of feeling, knowing, and acting in relation to these ordinary ... "objects," they play to and help us complete the formation of a historically specific kind of aesthetic subject: "us." (232)

Aesthetics is not about seeking after timeless, placeless judgments.

3. Aesthetic judgments are evaluations.

I make this point to further my larger points about how our particular sensibilities—that is, about how the ongoing summation of our aesthetic judgments—teach us how to live in a world together. The aesthetic judgments so far named—the cute, the beautiful, the ridiculous, the zany, the pretty, the interesting, the elegant, the ugly—are all about how we value the objects so judged and so how we regard and use them alone and together; the consequences of such judgments are present in the words I quoted earlier from Bourdieu, about how these judgments shape classes and the class judgments of others, but such consequences are also present in the results of the implicit attitude tests I earlier described and in the judgments that different cultures make about each other. We cannot separate our aesthetic responses from our judgments, and vice versa: to explore and question such judgments requires that we explore and question how and why we have the aesthetic feelings we do.

4. All texts are aesthetic.

From all I have been writing in these pages I hope you will take away that any text that engages us sensuously is aesthetic, by definition—and that all texts engage in teaching us what our senses ought to be. Hence all texts are aesthetic and worthy of being questioned for how they shape our engagements with the world and so each other: every text we or someone in our classes composes incorporates appeals that both request aesthetic response and shape how we sense and that therefore engage deeply with how we live together ethically.

* * *

If I have persuaded you that aesthetics is deeply ethical—in that it is about how we learn to be bodies—and that aesthetics is therefore about our relations with the world and with each other, then should not we be folding aesthetic considerations into our academic and professional writing classrooms as we help people in those classes understand how texts do their work? And if I have persuaded you that there are potential openings in digital times for reshaping our senses, or if you are already persuaded that the ways we've constructed digital environments participate in not only what we take to be our usual five senses but also in our senses of time, distance, identity, and connection, then should not we be folding aesthetic considerations into our academic and professional writing classrooms?

The articles by Ott and Keeling and by Gregory Clark that I mentioned in my introduction remark tensions between rhetoric and aesthetics: they find tensions in how we have tried to hold the rational, discursive, and cognitive apart from sensuous experience. Much current rhetorical theorizing does now concern this tension and does concern the rhetoricality of affect and embodiment. As a modest offering toward pulling that theorizing still more into writing classrooms, I list below a heuristic that might help students do the work of learning how not to separate our bodily feelings and judgments from our cognition and thought. The heuristic offers questions to use individually or in any combination, in whole or in part, to supplement other questions you might use in helping students learn rhetorically about their bodies, their texts, and the other materials of our shared worlds.

Preliminary Heuristic for Engaging Rhetorically with the Aesthetic Aspects of Texts

These questions can be asked to analyze a text made by others, or they can be asked of a text being produced.

- With what senses does this text engage? Does it focus you on *seeing* the text or objects, or on *hearing*, or on *touch*, or on . . . ? With what senses does this text *not* engage? Why do you think the text, given its purposes, engages with the senses it does?

- For each sense that a text engages, how does the text shape that sense toward its object? What judgments of taste/aesthetics are encouraged by the text? For example, if the text relies primarily on sight, does it ask you to look generously at what it puts before your eyes, or does it ask you to look critically or harshly or gently or lovingly? If the text relies on hearing, does the sound pull you close to what is being discussed or considered, or does the text push you back? If the text relies on you touching a screen or using a controller of some kind, is the touch asked of you gentle or commanding—or is it a slow touch or so fast that you have to learn not to think too much about how you must use your touch? How does the text engage your senses in these ways and encourage these ways of responding?

- If more than one sense is engaged by this text, how does the text relate the senses it engages? Does the text make one sense more important than others? For example, is there a soundtrack that pulls at you more than the visuals? Is there visual engagement that distracts you from how you are sitting or standing or moving? Or does the text ask you to lose sense of your senses? Why might the text prioritize the senses it does, in the way that it does?

- How do your sensory engagements with the text ask you to think about, evaluate, or otherwise judge the objects or the people within the text? Why might the text support these judgments, given its larger purposes?

- Does the text ask you to reflect on how your senses are engaged? Why or why not does it encourage reflection? How would the overall effect of this text change were it to ask you to reflect on the senses it engages and how it engages them?

- How does the text engage its audience's senses of time and space? That is, is the text set up to make you feel the world is passing by quickly or slowly? Does the text make you feel that you live inside your own head, inside a computer screen, or out in a field somewhere? How does the text achieve these effects? Why might the text seek these effects?

- How does the physical stuff or matter of the text shape an audience's responses? For example, is the text constructed out of stuff that feels good in your hands—or awful? How would the text be different were it to be made of other materials?

- How might this text be different were it to engage with other senses? Sketch out a redesign of the text so that it addresses other senses. For example, redesign a brochure as a podcast or a video game as a choose-your-own-adventure in print. How does the change in sensuous engagement shift how audience(s) are likely to respond to the text? How does the change in sensuous engagement shift how audience(s) are likely to feel about their own bodies, the bodies of others, or things in the world?

- How does this text encourage you to live with your own body? For example, when you finish with the text, do your eyes hurt? Do you feel like dancing? Do you feel happy and ready to move, or do you want to slump into your chair? Does the text encourage you to forget that you are a body? What happens in the text to encourage these feelings? How do these feelings contribute to the overall purposes of the text? How do these feelings contribute to how you feel about other humans and about nonhuman objects?

- How does this text encourage its audience to think about nonhuman things and objects? Does the text encourage its audience to think of things and objects as disposable or as having a life of their own (as animations sometimes do)? Given the text's purposes, why might it encourage its audience toward these attitudes?

- Find another text, from another time or place, that addresses similar issues to this one. How does your other text differ from this text in how it addresses senses? For example, does the other text look cluttered and so addresses sight as a sense that can (and that has the time to) distinguish among lots of information, or does the other text use much longer sentences, or does it represent people as more patient or . . . ?

- Imagine that this text was the only text from which people had to learn about how they should live in the world with each other and with things. What values and judgments would people make about living based on the text? (There's a *Star Trek* episode—"The Omega Glory"—about people who live with the preamble to United States Constitution as their "holy" text, without knowing any of its context; this episode suggests how a single text can shape larger behavior.)

The next questions relate particularly to the production of texts of all kinds.

- To help you achieve your purposes and develop the relations you hope to achieve with your audience(s), what sorts of sensuous engagements should your text shape with your audience(s)? Should your text encourage a focus on seeing what is happening or on hearing what is happening? Imagine your text addressing different senses in order to imagine what is possible.

- In your use of a photograph or video that focuses on people or objects, what values does the text assume your audience(s) will have about the people or objects? What values does the text encourage through its use of color, sound, motion, or placement of the people and objects relative to other people and objects? For example, are some people dressed in dark clothing who move heavily while others are dressed in light clothing who move brightly? Why does your text show different people differently?

- In your use of the sound of voices, how do different voices sound, and how does that shape how your audience(s) might respond to those voices?

I hope you judge these questions and this writing useful.

Notes

1. In addition to judgments about black and white folks, the Project Implicit website also allows one to check one's assumptions about gender and careers, skin tone, weight, age, Asian Americans, religion, weapons, Native Americans, Arabs and Muslims, disabilities, presidents, gender and science, and sexuality. (I list the categories as they appear on the website.)

2. The Wikipedia page titled "Sense" is as good an entrée as any into "nontraditional senses."

3. My visual support when I presented a version of this paper at the 2014 Maryland Conference included a long series of slides collected from a Google search for "teaching children the five senses." If you doubt the amount of repeated and widespread effort that goes into this particular construction of sensing in the United States and the West, do the search yourself, checking out not only the resulting Google links

to the "Web" (about 2 million results) but also the links to "Videos" (about 274,000 results) and "Images." The numbers certainly suggest that we as a culture have a lot at stake in reproducing this particular embodiment.

Works Cited

Ambady, Nalini, and Jamshed Bharucha. "Culture and the Brain." *Current Directions in Psychological Science*, vol. 18, no. 6, 2009, pp. 342–45.

Ames, Daniel L., and Susan T. Fiske. "Cultural Neuroscience." *Asian Journal of Social Psychology*, vol. 13, 2010, pp. 72–82.

Baumgarten, Alexander G. *Reflections on Poetry*. Translated by Karl Aschenbrenner and William B. Holther, U California P, 1954.

Bertrand, Marianne, and Sendhil Mullainathan. "Are Emily and Greg More Employable than Lakisha and Jamal? A Field Experiment on Labor Market Discrimination." *American Economic Review*, vol. 94, no. 4, 2004, pp. 991–1013.

Bierut, Michael, et al. *Looking Closer 2: Critical Writings on Graphic Design*. Allworth Press, 1997.

Bolter, Jay David, and Richard Grusin. *Remediation: Understanding New Media*. MIT P, 2000.

Bourdieu, Pierre. *Distinction: A Social Critique of the Judgment of Taste*. Translated by Richard Nice, Harvard UP, 1984.

Clark, Gregory. "'A Child Born of the Land': The Rhetorical Aesthetic of Hawaiian Song." *Rhetoric Society Quarterly*, vol. 42, no. 3, 2012, pp. 251–70.

Classen, Constance. *The Deepest Sense: A Cultural History of Touch*. U of Illinois P, 2012.

———. *Worlds of Sense: Exploring the Senses in History and across Cultures*. Routledge, 1993.

Drucker, Johanna. *The Visible Word: Experimental Typography and Modern Art, 1909–1923*. U of Chicago P, 1994.

Faigley, Lester. "Material Literacy and Visual Design." *Rhetorical Bodies: Toward a Material Rhetoric*, edited by Jack Selzer and Sharon Crowley, U of Wisconsin P, 1999, pp. 171–201.

Farnell, Brenda. *Do You See What I Mean? Plains Indian Sign Talk and the Embodiment of Action*. U of Texas P, 1995.

Foster, Hal, editor. *Vision and Visuality*. Bay Press, 1988.

Grau, Oliver. *Virtual Art: From Illusion to Immersion*. MIT P, 2003.

Hansen, Mark. *Bodies in Code: Interfaces with Digital Media*. Routledge, 2006.

Haverkamp, Michael. *Synesthetic Design: Handbook for a Multisensory Approach*. Translated by Michael Dudley, Birkhauser, 2013.

Hawhee, Deborah. "Rhetorics, Bodies, and Everyday Life." *Rhetoric Society Quarterly*, 36, 2006, pp. 155–64.

Holloway-Attaway, Lissa. "Beyond Representation: Embodied Expression and Social Media." *Digital Humanities Quarterly*, vol. 6, no. 2, 2012, www.digitalhumanities.org/dhq/vol/6/2/000118/000118.html#.

Howes, David, editor. *The Varieties of Sensory Experience: A Sourcebook in the Anthropology of the Senses*. U of Toronto P, 1991.

Howes, David, and Constance Classen. *Ways of Sensing: Understanding the Senses in Society*. Routledge, 2013.

Jay, Martin. *Downcast Eyes: The Denigration of Vision in Twentieth-Century French Thought*. U of California P, 1993.

Jütte, Robert. *A History of the Senses: From Antiquity to Cyberspace*. Translated by James Lynn, Polity, 2004.

Knight, Aimée. "Reclaiming Experience: The Aesthetic and Multimodal Composition." *Computers and Composition*, vol. 30, 2013, pp. 146–55.

Kress, Gunther, and Theo van Leeuwen. *Reading Images: The Grammar of Visual Design*. Routledge, 1996.

Lanham, Richard. *The Electronic Word: Democracy, Technology, and the Arts*. U of Chicago P, 1993.

Levin, David Michael, editor. *Modernity and the Hegemony of Vision*. U of California P, 1993.

Lindhé, Cecilia. "'A Visual Sense Is Born in the Fingertips': Towards a Digital Ekphrasis." *Digital Humanities Quarterly*, vol. 7, no. 1, 2013, www.digitalhumanities.org/dhq/vol/7/1/000161/000161.html.

Malnar, Joy Monice, and Frank Vodvarka. *Sensory Design*. U of Minnesota P, 2004.

Marx, Karl. "Economic and Philosophic Manuscripts of 1844." *Marxists.org*, 2009.

Moss-Racusin, Corinne A., et al. "Science Faculty's Subtle Gender Biases Favor Male Students." *Proceedings of the National Academy of Sciences of the United States of America*, vol. 109, no. 41, 2012, pp. 16474–79.

Munster, Anna. *Materializing New Media: Embodiment in Information Aesthetics*. UP of New England, 2006.

Ngai, Sianne. *Our Aesthetic Categories: Zany, Cute, Interesting*. Harvard UP, 2012.

Nosek, Brian A., et al. "Implicit Social Cognition: From Measures to Mechanisms." *Trends in Cognitive Sciences*, vol. 15, no. 4, 2011, pp. 152–59.

Ott, Brian L., and Diane Marie Keeling. "Cinema and Choric Connection: *Lost in Translation* as Sensual Experience." *Quarterly Journal of Speech*, vol. 97, no. 4, 2011, pp. 363–86.

Project Implicit. 2011, https://implicit.harvard.edu/implicit/index.jsp.

Rancière, Jacques. *The Politics of Aesthetics: The Distribution of the Sensible*. Translated by Gabriel Rockhill, Continuum, 2004.

Sacks, Oliver. *Seeing Voices: A Journey into the World of the Deaf*. U of California P, 1989.

"Sense." *Wikipedia*, last edited 13 Apr. 2018.

Spuybroek, Lars. "The Ages of Beauty: Revisiting Hartshorne's Diagram of Aesthetic Values." *Vital Beauty: Reclaiming Aesthetics in the Tangle of Technology and Nature*, edited by Joke Brouwer et al., V2 Publishing, 2012, pp. 33–60.

Starr, G. Gabrielle. *Feeling Beauty: The Neuroscience of Aesthetic Experience*. MIT P, 2013.

Virilio, Paul. *The Vision Machine*. Indiana UP, 1994.

Wysocki, Anne Frances. "Unfitting Beauties of Transducing Bodies." *Rhetorics and Technologies: New Directions in Writing and Communication*, edited by Stuart Selber, U of South Carolina P, 2010, pp. 94–112.

Section V

Two Perspectives on Assessment

14

It's Tagmemics *and* the Sex Pistols: Current Issues in Individual and Programmatic Writing Assessment

Kathleen Blake Yancey

The practice of writing is inherently a social act, while the assessment of writing has historically been constructed as an individual, almost asocial act; in that sense and in its prioritizing of efficiency, writing assessment has long been at odds with theories and practices of writing itself. More recently, however, writing assessment has developed its own theory and practice, both of which are more congruent with those of writing. Consequently, writing assessment doesn't have to be a misaligned afterthought to writing instruction; instead, it is positioned to play a central role in both curriculum and pedagogy. More specifically, as I highlight here, in many of its current features—from its attention to the writing construct locating our curricula and its potential to highlight multimodality, to its foregrounding of the importance of reflection in learning, and to its provision for the ways student writing exerts impact outside the classroom—writing assessment has taken its own social turn, contributing to its now-integral role in our curricula and pedagogy.

It's a truism that since the turn of the twentieth century—over a hundred years ago—we've witnessed two historical patterns in writing assessment. The first, what Michael Williamson has called "the worship of efficiency," takes the answer to a central question as the most important consideration in selecting or designing a writing assessment: "Which measure can do the best and fairest job of prediction with the least amount of work and the lowest cost?" As the last two factors in the question suggest, the question places emphasis not on how we help students or on which writings we might review but rather on factors of the "least amount of work" and "the lowest cost" and how they can contribute to a writing assessment. Given this question and the values it endorses, it's not surprising that many of our writing assessments are, in a word, disappointing—keyed not to evaluation or to assistance but rather to getting assessment-as-job done, and quickly at that.

The second historical pattern in writing assessment, outlined by Bill Condon, is what I call the pull of the past, that is, the tendency to tie promising and innovative assessment practices to the ghosts of the very assessments they intend to replace. Thus, the so-called new SAT test of writing, introduced in 2005, bases two-thirds of its score not on writing but on a multiple-choice test. The multiple-choice test thus does more than contribute to the "writing test": it anchors it and determines the score. Another example is closer to the academic home, a practice called "phase two portfolio scoring" (White). When portfolios are used for assessment, the entire portfolio is read and scored, as we might expect, but not so in phase two scoring: although students are asked to create a portfolio, that portfolio isn't read or scored. Instead, a reflective cover letter or essay is removed from the portfolio and scored independently, as though it were a stand-alone document. In phase two scoring, we thus revert to the assessment of choice that preceded portfolios and that portfolios were intended to replace: the holistically scored single essay (Yancey, "Looking Back"). In brief, progress in writing assessment is often shadowed by the pull of a past practice that tends to overwrite it.

And while not generally observed in the literature on writing assessment, it is equally accurate to point to a third historical pattern related to the first two but distinct from them, a pattern that I'll use to frame my remarks in this chapter: the overwhelming absence of social practices in our assessments.[1] Put simply, writers compose in social and material contexts, but we tend to assess their texts removed from such contexts, even in our more progressive models, which means that our definition of writing, or construct of writing, is at odds with our mechanism for assessing it.[2] If, however, we began *designing* writing assessments, both individual and social, with social practices as a primary consideration, we'd develop very different writing assessments, ones much more informed, useful, and ethical. It was the inclusion of social practices, for example, that was a big part of the draw for portfolios as they were first envisioned in the 1980s, when by design they included texts speaking to multiple contexts: traces of writing practices in multiple drafts and peer review responses, for example; evidence of a writer's ecology of writing through the inclusion of several genres; and inclusion of a writer's thinking through a reflective text articulating his or her analysis or interpretation of the portfolio contents and/or composer. Now, more commonly, portfolios are less oriented to such social practices and rather more oriented to outcomes and assessments. Put another way, the full if messy portfolios of yesterday tend today to be a set of finished products fronted by a mandated argumentative text in which a student is required to claim in terms of outcomes that he or she has met those outcomes—even if he or she hasn't. (And that's assuming that the reflection wasn't removed and assessed in isolation.) In this shift from messy to formalized, the social has been lost.

In addition, thinking in terms of social practices directs our attention to what we do *before* we assess writing, that is, to our reading of writing: to the ways we read different texts, for instance, from lyrical essays to research posters; to different kinds of texts, from a flyer in print to a multimodal essay in print; and to conventionalized readings and expectations common in different discourse communities and settings, be they first-year composition classes, writing across the curriculum, and writing in the disciplines programs; capstones and co-curricular experiences like internships and service-learning projects; or workplace, civic, and personal sites—in person, on paper, and on the web. Likewise, reflection itself is a social practice, both personal and social; thus, including reflection in assessment as a normal practice, regardless of the form that assessment takes, could help us review our own curricula and pedagogy so that we could improve. In addition to providing another learning opportunity to students, we would learn about various uses of reflection and the ways it can both support student writers and frame their writing for our reading. And not least, attending to social practices could help us include in any assessment model what Nadia Behizadeh calls the "impact factor," that is, what students have and are accomplishing with their writing, sometimes with our help—and often without it.

Taken together, thinking in terms of social practices can thus help us sort out issues in four areas that continue to vex writing assessment: (1) the construct of writing; (2) the ways we read, view, and listen to texts and thus assess them; (3) the role of reflection; and (4) the impact of students' writing.

And what about the key terms in the title of this chapter, the tagmemics and the Sex Pistols?

Tagmemics—a theory of language located in the three perspectives of particle, wave, and field developed by Richard Young, Alton Becker, and Kenneth Pike—never worked very well as a schema for invention, which was their hope, at least in part. But it's not quite the formalism that Geoff Sirc suggests, either, and as a tool for analysis, it is very helpful, providing a framework for understanding in which any phenomenon can be seen at the same time from three perspectives: as a single unit, as part of a wave, and as part of a field. Interestingly, this is the same point that Erika Lindemann made twenty years ago about the ways we teach writing: we can teach the product (a single unit); we can teach the process (a wave); we can teach the ecology (a field)—and we can teach all three at once. The history of writing assessment follows a parallel track. Historically, we have focused on the product, a single piece of writing decontextualized from its source, and that's the formalism that Sirc—and others like George Hillocks—deplores. We have, however, also assessed process,[3] sometimes assessing a single text in the context of the process pieces contributing to it, sometimes assessing a single text in the context of a stand-alone reflection, sometimes assessing folders of student work with finished texts and process pieces but no reflection. And as previously mentioned, we have assessed student writing—products and processes—participating in a fuller ecology, especially in portfolios; furthermore, when these portfolios include writing that students selected from outside our classes or programs, we begin to see an ecology unique to the student's own experience. Thus tagmemics, through emphasizing the potential of engaging any phenomenon from three perspectives at once, helps us see a text operating in several contexts.

And the Sex Pistols? In Sirc's article, he laments the loss of composition of a certain moment, what he sometimes calls Comp 77, but his larger point has to do with what the punk of the Sex Pistols represents: "Punk," he says, "forms a permanent theatre of tension—the dominant culture vs. the underground, the academic as against the innovative" (27), a theme not unlike assessment's worship of efficiency at odds with innovative practice. In addition, like the interests of Young, Becker, and Pike, Sirc's interest in innovation anticipates composing as a practice providing text for future composers; he highlights the remixing and repurposing practices increasingly common in composition classrooms in his attention to the Sex Pistols' graphic designer Jamie Reid's "re-making/re-modeling the materials of the dominant culture" for his texts publicizing the Sex Pistols (15). What's also interesting is Sirc's revision of standard criteria like goodness or weakness: "Whether they [the Sex Pistols] were good or not was irrelevant," Sirc quotes a fan saying; performers, and therefore composers, "are only as interesting as the emotions they generate, or the situations they catalyze" (16). Or: a kind of impact, in the terms Behizadeh specified.

My final prefatory note concerns what's excluded in this chapter's discussion about writing assessment. In the period spanning 1995–2015, we've seen a cottage industry of work in writing assessment. *College Composition and Communication* commissioned two review essays (Neal; Gallagher) on the topic in a five-year period (2010–2014) and published numerous articles on the topic

(Leaker and Ostman; Gere et al.), including a Braddock Award winner (Scott and Brannon); Bill Condon's review essay in a 2011 issue of *WPA: Writing Program Administration* included thirteen books. Given these texts—articles, review essays, and books—there are many topics within writing assessment to be addressed, some of which are neglected here, each important: students' languages, for example; translingualism, more generally; and machine scoring. Nor am I addressing how, since about 2000, the practices and beliefs in higher education have come under increasing scrutiny, criticism, and pressure, especially from the federal government (Yancey, "Writing Assessment"). Nor am I focusing on the funders and foundations, the textbook companies and testing agencies that likewise are increasingly and intrusively present in our institutional policies and our classrooms. All of that, and it is worrisome, is background for what I offer here: ways to conceptualize the role of social practices in writing assessment and, in the process, reinvent a progressive writing assessment keyed to the current moment. Put differently, the role of social practices underlies nearly all current issues in writing assessment; providing a way forward in this area will enhance both writing assessment and the development of student writers.

The Construct of Writing

The first issue vexing writing assessment is what is called the construct of writing. Put more accurately, the construct isn't the problem; instead, it's the way we have—but mostly have not—defined the construct. As it sounds, *construct of writing* refers to the model or theory of writing that underlies any assessment practice. For most of the twentieth century, the construct of writing was universal: there was a working assumption that good writing was good writing was good writing—regardless of task, genre, or rhetorical situation, regardless of site of composing, culture, or language. This assumption about writing as a generalized domain provided the foundation for assessments of both the multiple-choice variety and the holistically scored essay. In contrast, the assumption informing portfolios—even in the 1980s—was that writing isn't a single practice but rather is a set of localized and differentiated practices that vary in multiple ways, including by rhetorical situation and genre. Thus,

> the ability to write one kind of document does not automatically guarantee the ability to write another kind of document; the successful completion of a generic "research paper" does not ensure the successful completion of a journal article or business proposal or laboratory report. (Paretti and Powell 4)

Designed by teachers and not assessment experts, portfolios sought to sample a *range* of writing genres in an effort to tap a fuller representation, or construct, of writing. And it's worth noting that teachers used portfolios not out of any concern for a technical version of validity but largely because they understood that portfolios, as opposed to stand-alone essays, would help students demonstrate a much fuller picture of their writing, especially in the context of the classroom. With portfolios, then, we have a mechanism for assessing a fuller construct of writing.

As we know, however, print isn't the only medium hosting portfolios: we have electronic portfolios. In 1994, over twenty years ago, the National Council of Teachers of English hosted a conference on electronic portfolios; in 1996, *Computers and Composition* offered a special issue on electronic portfolios. Still, even today the construct of writing continues to be predicated on a print model; we haven't theorized a construct of writing that includes print, electronic, and networked texts.[4] Moreover, even our writing construct located in print is impoverished: what we assess are the words; we don't account for the many other dimensions of writing, what Gunther Kress calls modes, that allow us to make meaning—font style and size, for example, and page layout, BOLDFACING, *italics*, and underlining These modes are especially available with digital technology, of course, but multimodality has always characterized writing, regardless of medium, as Lester Faigley explains:

> Images and words have long coexisted on the printed page and in manuscripts, but relatively few people possessed the resources to exploit the rhetorical potential of images combined with words. My argument is that literacy has *always* been a material, multimedia construct, even though we only now are becoming aware of this multidimensionality and materiality because computer technologies have made it possible for many people to produce and publish multimedia presentations. (175–76)

In the assessment world, instead of seeing texts as multimodal and rewarding adept writers who use multiple modalities to make meaning, and helping students who don't know yet how to do that, we assess words as the single semiotic resource that composers tap, words as writing. They contribute to writing, of course; but if there is a surface on which writing appears—be it cave wall or animal hide, cell phone screen or construction paper—there is the beginning of more than words and thus more than words with which we can make meaning. Assessing only words in writing assessments is a problem in that such a word-centered construct of writing misrepresents composing. It's a problem for students who—required to compose in Courier twelve-point font within one-inch margins—are *told* which modes to use instead of being taught about a suite of modalities available for making meaning and the ways to use them. It's a problem for any assessment aiming for consequential validity, which is an assessment-informed way of saying any assessment that intends to help students. Not to put too fine a point on it: if our construct of writing is impoverished, so too our curriculum, our pedagogy, and—not least—our students.

Some classrooms and programs are representing writing more fully, although most imagine multimodality as a feature of digital texts, not of print, when in fact it's a feature of both textualities. Still, identifying criteria keyed to what we value in multimodal texts is a good start, an exercise that Emily Wierszewski's research helps us understand. Wierszewski studied how faculty create their own criteria for multimodal texts: they develop a new lexicon—with terms like creativity and movement—and thus a (Burkean) new way of seeing. But such research is both infrequent and nascent; too often we rush to a rubric, much like a student who wants to outline before inventing, before identifying what it is that she wants to say. In our case, rubric in hand, we assess new texts much the same way we assessed the old.

As an aside: Recently I've been trying to devise an approach to assessing electronic portfolios that fosters less evaluation and more conversation (Yancey, "Social Life"). In thinking back on my own practice, I can't remember a scoring guide leading to a conversation; instead, it truncates or even precludes any social or intellectual exchange.

And a question we all need to answer: What is the construct of writing our classes and our programs enact, and which of the social practices that contextualize and constitute writing are we including?

Reading, Interacting with, and Reviewing Texts

The second issue vexing writing assessment: How do we read, interact with, and review texts? In terms of print texts, the field has developed some very helpful research. Judy Pula and Brian Huot, for example, studied the role of using a rubric in forming teachers' response strategies. Somewhat counterintuitively, they found that teachers who learned how to apply a scoring guide were more, rather than less, inventive in their response to student work; simply having a schema to tap as they responded to student work seemed to free them up to read the student text more capaciously. In a later study of how faculty read print portfolios, Bill Condon and Liz Hamp-Lyons found that readers rushed to judgment, although in different ways:

> We have found again and again in portfolios of different kinds, at
> different times, from different readers, a clear suggestion that readers
> do not attend equally to the entire portfolio. . . . Readers seemed to
> go through a process of seeking a "center of gravity" and then read
> for confirmation or contradiction of that sense. (182)

And we have some thinking about how we read online. Theorizing such
practices, for example, Jim Sosnoski identified six different kinds of reading,
a set of practices constituting what he calls "hyper-reading": "Understanding
that when I use the expression 'hyper-reading' . . . I refer to its 'constructive'
aspects, we can say that it differs from reading printed texts or expository
hypertexts in several ways" (163). More specifically, hyper-reading, he notes,
is characterized by the following activities:

1. filtering: a higher degree of selectivity in reading (and therefore)
2. skimming: less text actually read
3. pecking: a less linear sequencing of passages read
4. imposing: less contextualization derived from the text and more from
   readerly intention
5. filming—the ". . . but I saw the film" response, which implies that
   significant meaning is derived more from graphical elements as from
   verbal elements of the text
6. trespassing: loosening of textual boundaries
7. de-authorizing: lessening sense of authorship and authorly intention
8. fragmenting: breaking texts into notes rather than regarding them as
   essays, articles, or books. (163)

And there are other kinds of theories about reading that emerge from a
documented, sustained engagement with a single text. As part of a project
intended to develop a reading process specific to electronic portfolios, initially
developed on-screen, Stephen McElroy, Elizabeth Powers and I considered
a single electronic portfolio so as to identify the reading practices we called
"viewing/reading":

> Beginning to review the portfolio, we first decided, each of us sepa-
> rately, which page to click first, then which link to click second. . . .
> Upon encountering a text, we [also] needed to decide what to *do* with
> it. Would we, for example, click the contact screen and complete the
> email form so that we were both reading *and* writing? Would we
> download print texts—which ranged from the one-page resume to
> the multi-page research project—to our computers and read those,

and if so, would we read them through completely and carefully, or would we skim them . . . ? Would we link to a video and not read it, but rather watch it? (Yancey et al.)

At the end of this process, we identified three kinds of viewing/reading:

First, there's the viewing/reading of each individual text—which itself involves different reading practices for different kinds of texts—print, static screen, animated multi-media (animated video files, academic "papers," etc.). Second, there's the reading of the portfolio on the screen, where basically one toggles from the reading of the screens and print files and animated files to the reading of the portfolio as a composition, and where in this toggling one constructs the portfolio one is viewing/ reading. And third, there's a spatial reading, which helps us understand in practical, embodied, and theoretical ways the portfolio as a composition.

In other words, what we theorize is a set of reading and viewing practices operating simultaneously and at different levels of scale: this is the reading of the future that we are engaged in now. These are also the kinds of viewing/ reading that we have to build into any assessment process addressing electronic portfolios or even new media texts with more than one kind of file format.

Still, little of our research to date has been designed to help us read or view, read/view, or assess other kinds of multimodal texts. How, for example, might we read the *New York Times*' publication "Snow Fall: The Avalanche at Tunnel Creek"? "Snow Fall," John Branch's account of backcountry skiing, a dangerous kind of skiing off of sanctioned ski trails, depicts a skiing adventure gone wrong, resulting in the deaths of several skiers. Somewhat like an electronic portfolio but on steroids, "Snow Fall" incorporates various kinds of texts—among them, animations of weather patterns, of ski paths, and of the avalanche itself; graphics, written prose, and still photos; and video interviews—integrated into a single composition presented through a dynamic interface, the full set of texts contributing to an analytical narrative characterized by *Wired* magazine as an "experience-based feature."

Or we might consider Aisha Harris and Chris Wade's "Cool Moves: From Al Jolson to Channing Tatum; How Dancing in the Movies Creates—and Commodifies—Cool," a ten-minute video published by *Slate* in 2013. As its sub-subtitle and synopsis suggest, the video—composed principally of movie clips of dancing with their accompanying music scores, including a diverse range of clips and dancers (for example, Frank Astaire, Audrey Hepburn, the dancers of *West Side Story*, and John Travolta in *Saturday Night Fever* and *Pulp*

Fiction), and a continuous voice-over—is basically an academic thesis-and-support analysis on video. It's enlivened by the film clips. It's a treat to remember the films or to encounter famed actors for the first time along with the music. It's more than a treat, however; it's nearly a seduction.

As my description attests, I'm particularly interested in the coolness video. Using this video as a thought experiment, I gave a workshop for graduate students at another university on ways to assess digital multimodal texts. I shared this video, intending to think about how we might assess such texts more generally. Borrowing Richard Lanham's well-known schema from *The Electronic Word*, I suggested that we look at the video in a tagmemic fashion, through two lenses simultaneously, in this case using Lanham's lenses of looking at and looking through, as explained by Heidi McKee:

> Lanham argues that as printing techniques evolved, the physical, constructed nature of printed texts took on the goal of unselfconscious transparency—to convey "unintermediated thought, or at least what seemed like unintermediated thought" (4). Rather than notice and question, say, the 8½ 11-inch paper and Times New Roman typeface, traditional print media (to be judged successful) often positioned readers to read through the structure of the text to the message being conveyed. But as Lanham repeatedly asserts, all text is, of course, mediated, and it is important for people to recognize the mediated nature of texts. Such a recognition is made easier by computers, Lanham claims, because with computers people can create malleable, interactive digital texts in which "the textual surface has become permanently bi-stable. We are always looking first AT it and then THROUGH it, and this oscillation creates a different implied ideal of decorum, both stylistic and behavioral. Look THROUGH a text and you are in the familiar world of the Newtonian interlude. . . . Look AT a text, however, and we have deconstructed the Newtonian world" (5). (McKee 118)

In other words, especially in reviewing electronic texts, we look twice: AT and THROUGH. My sense of this video is that the looking AT—and also the listening to, as I explain below—overwrites the looking THROUGH. It's sufficiently fun to look at the video's dancing and listen to its music that we don't seriously consider the material being discussed: we don't look THROUGH it to consider its claims and evidence. We don't ask, Is this in fact the way that movies represented cool? What is cool? Is there any kind of counternarrative to this argument that might be included? The students in the workshop, as had I on first viewing the video, found much to like—and indeed, in their case, nothing to critique. Or, rather, the combined magic of screen and music requires a new approach to writing assessment.

In thinking about this issue as instantiated in the coolness video, I've widened the interacting tagmemic-like lenses—for this text, it's looking at/looking through *and* listening to. The music is sufficiently powerful that viewers tend to tap their feet and nod their heads—if not silently sing along—and as Kenneth Burke reminds us, our participation in an action functions as a kind of endorsement. Likewise, one could imagine adding other lenses to accommodate other modes—such as, say, the proverbial Jody Shipka text-on-ballet-slippers. But for this text, this video, these three lenses, working together—looking at/looking through/listening to—provide a useful schema for us to think about how the central modalities of the video work, or not. And a colleague of mine, Jacob Craig, expands on such looking at/looking through/listening to in his assessment of the coolness video (expressed to me in personal communication):

> I do think the draw of the video is the music. And I don't like that she [the narrator in the voice-over] talked over the music. More often than not, the dances she discusses feel secondary to the music itself. Dance plays a supporting role; music carries the lion share of what these pieces are trying to do. I think her argument would have been better served by providing more opportunity to look at and to listen to.

I agree with Jacob: assessing this text calls for a very different kind of reading/ viewing practice, one that brings multiple lenses, tagmemic-like, together. When one is looking through, the video focuses on a kind of cool but defines it in an uncool, sanitized way; that's a looking-through problem. When one is listening to, the voice-over is too much, too neat, too clean. When one is looking at, the video is smart—and nostalgic for those of a certain age; it carries the day but undermines the overall effect. And the three lenses don't quite work together, aren't sufficiently tagmemic. The focus, the looking through, is coolness, a Sex Pistols topic, but without any of the punk energy: it's thesis and support all over, at odds ironically with the very looking-at topic of coolness. My larger point: in this kind of discussion, we are coming to the terms that can help us assess multimodal texts.

As an aside: To the extent that we are assigning this kind of text and beginning this kind of assessment, machine scoring—the review of a text by a computer keyed to items like kinds and number of vocabulary and sentences—is not an option; it's the very antithesis of a writing assessment incorporating social practices. Moreover, if *all* texts are by definition multimodal—the electronic portfolio of print texts and still images composed in first-year composition; the poster in writing across the curriculum; the geographically team-based texts incorporating graphics and mathematical analysis seamlessly in writing in the disciplines (as in the Mechanical Engineering Writing Plan at the University of Minnesota); and the résumé in a capstone—then machine scoring is never a viable option. Thus, as we design assessments congruent with the multimodality that texts display, ones with a writing construct informed by multimodality, we also show how and why "robograding" is inadequate for assessing them.

Reflection

A third area vexing writing assessment is reflection, and like the writing construct, the problem isn't with reflection itself but with how we do and don't define, assign, support, and assess reflection. Having talked to many faculty on several campuses about reflection and having reviewed numerous portfolios, I'm aware that we have multiple definitions—ranging from metacognition, account of process, and self-assessment to synthesis, rhetorical explanation, and exploration. Furthermore, during the last fifty years in writing studies, reflection has shifted, from our defining reflection as an innate component of composing per Sharon Pianko, to defining it as a secondary text to a portfolio's primary texts per Chris Anson, to defining it as a design explanation for digitally multimodal texts per my colleague Michael Neal (Yancey, "Contextualizing Reflection").

In this last sense, Neal's approach is in the spirit of a reflective text that accompanied the *New York Times*' photo essay "On the Ground in Israel and Gaza," a visual essay "captur[ing] scenes from the most recent outbreak of war" in the summer of 2014. In commenting on the photo essay created by Paolo Pellegrin and Peter van Agtmael, editor Jake Silverstein contextualizes it, beginning with the description of a portfolio-like process of collecting, selecting, and assembling so that the reader/viewer understands the way it was composed:

> A photo essay, like anything else we publish, involves editing—a process of selection that distills the large number of images we receive from the photographers into a narrative composition. We look for strong individual images that work well together, coalescing into a story that represents the experience of the photographers in the field. Once we settled on a set of images, we began to explore different ways of presenting them. We eventually decided to intersperse them, as a way of emphasizing that the fates of average Israelis and Palestinians are intertwined.

Likewise, the editor comments on what the composition cannot achieve: "Peter was not allowed to embed with the Israeli military, though he still managed to produce a few shots of soldiers near the border. Paolo, like most photographers in Gaza, was not able to take pictures of Hamas fighters." Here, Silverstein, anticipating the needs of a reader/viewer and the (social) practice of reading/viewing, and thinking about the project and about the composition of the project, provides relevant information, in particular about the constraints governing the project. That's one version of reflection especially keyed to the needs of a reader/viewer, especially for a project located on the web.

Inside the academy, there is another promising reflective practice. Jeff Sommers, for instance, has used reflection to help students tease out their beliefs about writing as they begin a course, to trace how those beliefs change, and don't, so that by the end of the term students have articulated both what they as individuals believe and what they as a community believe. It's reflection as individual practice in a social context allowing students to relativize their own beliefs. Similarly, working with teams under the auspices of the Inter/national Coalition for Electronic Portfolio Research, colleagues and I have explored the distinctive contribution that reflection makes to the learning showcased in, and the assessment of, electronic portfolios. What some coalition members, at least, have concluded is that there is a kind of meaning-making operating at the intersection of the personal and the academic that can only be made through a process of reflection. In this view, reflection is not secondary or complementary but is an independent domain—its own construct.[5] And if so, what we need is a curriculum in reflection: one with readings and activities engaging students in reiterative practice and supporting them—returning over the course of a term to the same questions, for example—toward the goal of making knowledge about their own composing, one that also helps them understand why they would want to do so.

As an aside: Too often we ask students to practice without introducing them to and helping them understand the theory of the practice. A curriculum in reflection has as its intent both; it is this kind of curriculum that helps students articulate what they are learning in order to secure that learning and to understand the value of doing so, an approach that is also being employed in policy documents like the most recent WPA Outcomes Statement.[6]

The Role of Impact

The final issue vexing writing assessment is impact. Stated simply, if students were invited to document the impact of their own writing, there would be no need for additional writing assessment. Writing assessment inquires into questions about students' writing—what genres they can write in, how they have developed as writers, and, mostly, how well they can write. But if students demonstrated how well they write by pointing to the impact of their writing, we'd have a better, more "natural" assessment than any we contrive.

A version of this approach has been proposed by literacy assessment scholar Nadia Behizadeh. In "Mitigating the Dangers of a Single Story: Creating Large-Scale Assessments Aligned with Sociocultural Theory," Behizadeh suggests that in their reflections, K–12 students should include an "evaluation of the impact of their writing. Students will evaluate if their writing has the intended effect on their audience, which is important for the social purpose of writing." Given the kinds of writing postsecondary students do now—in service-learning contexts, on the web, in their own lives, on Wikipedia, in newsletters and zines, in the journal *Young Scholars in Writing*—asking them to include the texts themselves is just as important as reflecting on and evaluating them. Taking a note from Sirc, we would find it useful to ask, not only what they have learned in the composing and sharing of a text, but also to ask what emotions their writing generated and what actions it catalyzed. We talk about writing making a difference in the world—as we do in our themes for the annual meeting of the Conference on College Composition and Communication—"Connecting the Text and the Street," for example, "Making Composition Matter: Students, Citizens, Institutions, Advocacy," and "The Public Work of Composition"—but when it comes time for assessment, we tend to revert to the narrow, to limit our gaze to the school context, to standard criteria. When we do that, much is lost: the role of the social in writing is minimized if not lost altogether; we lose the opportunity to learn from students what difference, if any, they have made and how that happened. Designing impact as part of courses and assessment may, of course, make all of us a bit uncomfortable at first. Students may feel that the bar has been raised, and unfairly. Teachers may experience anxiety at what would be a loss of control. But, ultimately, the purpose of writing *is* to have an impact. A good beginning point might be to consider how the term *impact* would be defined, what it might look like in terms of specific texts, and how we could forward this agenda.

And as an aside, let me emphasize the role of the world in impact. As is clear, many of my examples—the coolness video from *Slate*; "Snow Fall" and "On the Ground in Israel and Gaza" from the *New York Times*—come not from school but from the world. That was a deliberate choice. It's not that students aren't making these kinds of texts; it's that when we locate our own practices in world practices, we create a connection helpful to us all. We—faculty and students—are part of the world, its values, its beliefs, its practices; contextualizing our practices in terms of the world's points us away from educational measurement and government metrics, away from any of the many, many companies that will aggregate results and analyze and interpret them and tell us what they mean and that will continue to drain writing assessment of the very social practices that make writing meaningful, those of writing, reading, and assessment as shared practices, as we look at, look through, talk to, and think with.

A Concluding Scene

In a first-year composition class, students begin by identifying their beliefs about writing both individually and collectively, providing a starting point for the class. When they review the syllabus, they are surprised to find that they are supposed to design multimodal texts of their choice. To prepare, they engage in another unfamiliar task, considering what they want the impact of that text to be: the text is expected to accomplish something. Questions scaffolding their progress are introduced: Who will read the text if it's print or view/read it if it's electronic? And to what end? What do you want the text to accomplish?

Later in the term, students are researching their topics, talking about planned impacts with their colleagues, writing reflective planning memos. The texts drafted, students engage in a peer review process keyed to a heuristic,[7] one oriented to looking at / looking through:

1. Who is the composer?
2. What does he/she do well in this text?
3. What questions are raised (that could be addressed in a revision)?
4. What do you see looking through?
5. What do you see looking at?
6. How/do they match?
7. If this were my text, I'd _______________________________ .
8. Other

After peer review, students revise and share again, especially with the audience they expect to impact. As part of the assignment, the students document that impact and reflect upon it: Did the text have the anticipated impact, and if so, how? Or if not, how not? What in the process was learned about composing? Texts like these contribute to an electronic portfolio that is read by the teacher using a looking through / looking at framework;

some electronic portfolios are collected and used for program assessment. For this latter activity, the most important task is twofold: ascertaining the *collective impact* of the students' texts and learning from the students what they have learned about how to create impact and what that means for their future composing.

This scenario operates within Vygotsky's zone of proximal development, which means it's possible. To enact this scenario, we would need

- to define a writing construct as always multimodal, including both print and digital;
- to read using a tagmemic, Lanham-like approach located in concurrent looking at / looking through / listening to lenses;
- to design reflection that is a theoretically based practice; and
- to orient our assignments to impact à la Sirc.

We can do all this; moreover, as we do, we'll find that writing assessment, like the composing and texts it attends to, can be engaging, meaningful, and helpful.

Notes

1. Social practices, of course, have typically been deliberately excluded from writing assessment, as we see in large-scale assessments where students are forbidden to engage in any social activity.

2. There are exceptions to this rule: see, for example, Robertson's account of a placement exercise in which students do read, talk, and share as part of the assessment.

3. See, for example, Faigley and coauthors' *Assessing Writers' Knowledge*.

4. An exception to this generalization can be found in a gateway course titled Writing and Editing in Print and Online that is part of Florida State University's major in editing, writing, and media. The course requires that students compose both in and across three spaces: print, screen, and network.

5. Other curricula provide new questions for writing assessment: For example, in first-year composition courses keyed to writing about writing (WAW) and teaching for transfer (TFT), what is also salient is what writing knowledge the student has acquired. How we assess for that component of a writing education has yet to be addressed.

6. In the new WPA Outcomes Statement, several of the changes spoke to this point precisely: students need to know not only how to compose but why to maintain, for example, certain conventions. See Dryer and coauthors.

7. I have used this heuristic in two versions of an electronic portfolio class, and students seem to find the identification and separation of the looking at / looking through schema helpful. They also appreciate the query about the match of the two: that is, how they work together or do not—which of course speaks to the text in question.

Works Cited

Behizadeh, Nadia. "Mitigating the Dangers of a Single Story: Creating Large-Scale Writing Assessments Aligned with Sociocultural Theory." *Educational Researcher*, vol. 43, no. 2, 2014, pp. 125–36.

Branch, John. "Snow Fall: The Avalanche at Tunney Creek." *The New York Times*, 20 Dec. 2012, http://www.nytimes.com/projects/2012/snow-fall/index.html#/?part=tunnel-creek.

Condon, William. "Review Essay—Reinventing Writing Assessment: How the Conversation Is Shifting." *WPA: Writing Program Administration*, vol. 43, no. 2, 2011, pp. 162–82.

Dryer, Dylan B., et al. "Revising FYC Outcomes for a Multimodal, Digitally Composed World: The WPA Outcomes Statement for First-Year Composition (Version 3.0)." *WPA: Writing Program Administration*, vol. 38, no. 1, 2014, pp. 129–43.

Faigley, Lester. "Material Literacy and Visual Design." *Rhetorical Bodies*, edited by Jack Selzer and Sharon Crowley, U of Wisconsin P, 1999, pp. 171–201.

Faigley, Lester, et al. *Assessing Writers' Knowledge and Processes of Composing*. Ablex, 1985.

Gallagher, Chris. "Review Essay: All Writing Assessment Is Local." *College Composition and Communication*, vol. 65, no. 3, 2014, pp. 486–505.

Gere, Anne, et al. "Local Assessment: Using Genre Studies to Validate Directed Self-Placement." *College Composition and Communication*, vol. 64, no. 4, 2013, pp. 605–33.

Hamp-Lyons, Liz, and William Condon. "Questioning Assumptions about Portfolio-Based Assessment." *College Composition and Composition*, vol. 44, no. 2, 1993, pp. 176–90.

Harris, Aisha, and Chris Wade. "Cool Moves." *Slate*, 18 Oct. 2013, http://www.slate.com/articles/life/cool_story/2013/10/how_movies_express_cool_through_music_and_dance_video.html.

Hillocks, George, Jr. *The Testing Trap: How State Writing Assessments Control Learning*. Teachers College P, 2002.

Kress, Gunther. *Multimodality: A Social Semiotic Approach to Contemporary Communication*. Routledge, 2010.

Lanham, Richard. *The Electronic Word*. U of Chicago P, 1993.

Leaker, Cathy, and Heather Ostman. "Composing Knowledge: Writing, Rhetoric, and Reflection in Prior Learning Assessment." *College Composition and Communication*, vol. 61, no. 4, 2010, pp. 691–717.

Lindemann, Erika. "Three Views of English 101." *College English*, vol. 57, no. 3, 1995, pp. 287–302.

McKee, Heidi. "Richard Lanham's *The Electronic Word* and AT/THROUGH Oscillations." *Pedagogy: Critical Approaches to Teaching Literature, Language, Composition, and Culture*, vol. 5, no. 1, 2005, pp. 117–29.

Mechanical Engineering Writing Plan. University of Minnesota, undergrad.umn.edu/writing-enriched-curriculum-writing-plans.

Neal, Michael. "Review Essay: Assessment in the Service of Learning." *College Composition and Communication*, vol. 61, no. 4, 2010, pp. 746–58.

Paretti, Marie C., and Katrina M. Powell, editors. *Assessment of Writing*. Association for Institutional Research, 2009.

Pula, Judith J., and Brian Huot. "A Model of Background Influences on Holistic Raters." *Validating Holistic Scoring for Writing Assessment*, edited by Michael Williamson and Huot, Hampton Press, 1993, pp. 237–65.

Robertson, Alice. "Teach, Not Test: A Look at a New Writing Placement Procedure." *WPA: Writing Program Administration*, vol. 18, nos. 1–2, 1994, pp. 56–63.

Silverstein, Jake. "Editor's Note: How We Thought about Our Essay on Israel and Gaza." *The New York Times*, *The 6th Floor* blog, 28 Aug. 2014, 6thfloor.blogs.nytimes.com/category/photography-2/.

Sirc, Geoffrey. "Never Mind the Tagmemics, Where's the Sex Pistols?" *College Composition and Communication*, vol. 48, no. 1, 1997, pp. 9–29.

Scott, Tony, and Lil Brannon. "Democracy, Struggle, and the Praxis of Assessment. *College Composition and Communication*, vol. 65, no. 2, 2013, pp. 273–98.

Sommers, Jeff. "Reflection Revisited: The Class Collage." *Journal of Basic Writing*, vol. 30, no. 1, 2011, pp. 99–129.

Sosnoski, Jim. "Hyper-Readers and Their Reading Engines." *Passions, Politics, and 21st Century Technologies*, edited by Gail Hawisher and Cynthia Selfe, Utah State UP / National Council of Teachers of English, 1999, pp. 161–77.

White, Edward M. "The Scoring of Writing Portfolios: Phase 2." *College Composition and Communication*, vol. 56, no. 4, 2005, pp. 581–600.

Wierszewski, Emily. "'Something Old, Something New': Evaluative Criteria in Teachers' Responses to Student Multimodal Texts." *Digital Writing Assessment and Evaluation*, edited by Dànielle DeVoss and Heidi McKee, Computers and Composition Digital Press / Utah State UP, 2013. E-book unpaginated.

Williamson, Michael. "The Worship of Efficiency: Untangling Theoretical and Practical Considerations in Writing Assessment." *Assessing Writing*, vol. 1, no. 2, 1994, 147–74.

Yancey, Kathleen Blake. "Contextualizing Reflection." *A Rhetoric of Reflection*, edited by Yancey, Utah State UP, in press.

———. "Looking Back as We Look Forward: Historicizing Writing Assessment." *College Composition and Communication*, vol. 50, no. 3, 1999, pp. 483–503.

———. "The Social Life of Reflection: Notes toward an ePortfolio-Based Model of Reflection." *Teaching Reflective Learning in Higher Education*, edited by Mary Elizabeth Ryan, Springer, 2014, pp. 189–203.

———. "Writing Assessment in the 21st Century: A Primer." *Exploring Composition Studies: Sites, Issues, Perspectives*, edited by Kelly Ritter and Paul Kei Matsuda, Utah State UP, 2012, pp. 167–87.

Yancey, Kathleen Blake, et al. "Composing, Networks, and Electronic Portfolios: Notes toward a Theory of Assessing ePortfolios." *Digital Writing Assessment and Evaluation*, edited by Dànielle DeVoss and Heidi McKee, Computers and Composition Digital Press / Utah State UP, 2013. E-book unpaginated.

Young, Richard, et al. *Rhetoric: Discovery and Change*. Harcourt, 1970.

15

What If the Common Core Standards Actually Work? For College Writing, Partly Cloudy with a Chance of Rain

Doug Hesse

Many authors in this volume have discussed revisions of college writing programs, curricula, and pedagogies. The success of any change is considerably affected by the experiences, knowledge, skills, and attitudes of students before they matriculate. This chapter revisits some of the early critical stages in the evolution of the Common Core Standards and analyzes how K–12 students will experience reading and writing under an initiative that has mixed implications for college writing instruction.

A few times in my professional life, I've been in the presence of obvious power. One of them was 2002, in Chicago, in a suite in the Palmer House Hilton. Various officers from the Conference on College Composition and Communication (CCCC) and the National Council of Teachers of English (NCTE) were there, among them Shirley Logan, Kathi Yancey, and Anne Gere. We were meeting Gaston Caperton, a two-term former governor of West Virginia and then president of the College Board. In collaboration with the Business Roundtable and a newly constructed enterprise called the National Commission on Writing, the College Board had just drafted *The Neglected "R": The Need for a Writing Revolution*. NCTE leaders had received embargoed copies, and we were being asked to respond to the draft, perhaps endorse it. We registered several reservations that complicated the view of writing in the report. Caperton listened politely, even warmly, but ended the meeting by noting that they'd already lined up sports anchor Bob Costas as national spokesman, so the effort was heading forward with or without us. A year later the group produced a second report, *Writing: A Ticket to Work . . . or a Ticket Out*. Not much seems to have come of these projects, mostly because they failed to get buy-in from the teachers needed to implement them. However, I suggest that they lay dormant only to reemerge in a new guise.

Consider a second scene. It's the summer of 2009. The National Governors Association, the Chief State School Officers, and a group called Project Achieve have just drafted *The Common Core Standards for College and Career Readiness*. NCTE gets a copy in advance of its September 21, 2009, release for public commentary (National Governors), and I'm on a committee chaired by Randy Bomer (University of Texas and past president of NCTE) to write a response. We do. Later that fall I'm asked to represent NCTE and the Modern Language Association at a meeting in Washington, DC. There are a dozen or so English folks. Anne Gere (University of Michigan and also a past NCTE president) is our chair. Representatives of the Governors Association and the Chief State School Officers are there, too, along with a person unknown to most of us: David Coleman—who turns out to be the primary editor in charge of creating the Common Core. I end up sitting next to Coleman for the day. He's earnest and intense but friendly. We whisper back and forth, swapping notes. Toward the end of the day, as we're trying to formulate recommendations, Coleman makes a comment that catches the group by surprise. He shares that the standards writing group has progressed well beyond the draft we'd seen. In fact, they have "back-mapped" standards onto the whole curriculum, starting with kindergarten. I'm thinking that I must not have been paying attention, since these were presented as high-school-to-college standards, but when I look around the room, it's clear that everyone else is stunned, too. We're like the boy in the James Joyce story "Araby." We'd all gone to the bazaar expecting a grand experience only to confront our own vanity.

David Coleman, of course, has gone on to succeed Gaston Caperton as president of the College Board, and the Common Core has gone on to rival Obamacare for Tea Party vitriol, never mind its bipartisan, even conservative roots in state governments. In an October 2014 cover story, *Mother Jones* illustrated loathing for the Common Core with an image of children marching into the gaping maw of a huge wormy apple (Murphy). Proving yet again that politics makes strange bedfellows, many on the political left have rallied around complaints about excessive testing and a narrow curriculum. Coleman's candid but intemperate observation in 2011 that "people don't really give a sheet [*sic*] about what you feel or what you think" hasn't helped him. Nor, certainly, has the rampant testing-industrial complex that sieges teachers and public education. The privatized testing imperative has fueled progressive ire, as has the concern that the standards are problematic in terms of cognitive development and student motivation, shortsighted in casting education as job training for a narrowly imagined civic life, and injurious in failing to account for the varied lives of diverse students.

Of course, the reactionary critique wells from different sources, and it ought to give pause to progressives. It's striking, for example, when the Southern Poverty Law Center, that stalwart force against racism, hate crimes, and repression, sees fit to issue a report against the Common Core, as it did in May 2014. That report asserts that "by raising the specter of 'Obamacore,' activists on the radical right hope to gain leverage against their real target—public education itself" (4).

Given this fury from right and left, I'd like to pose a simple question: What if the Common Core actually works?

Overview of the Common Core Standards

The Common Core is a set of performance standards in mathematics and language arts for each grade level, along with illustrative materials and performance levels. It's not a required curriculum, the project's informational website contending that the standards "do not mandate the use of any particular curriculum" (Common, "About"). It's not a set of tests, the sponsors declaring that implementation "does not require data collection" and that "the means of assessing students . . . are up to the discretion of each state and are separate and unique from the Common Core" (Common, "Myths vs. Facts"). It's not even a required federal program, though decisions of the Obama administration that tied funding to it have clouded that point. All of the other points have been clouded, too, as the standards have prodded, through testing, the formation of a de facto curriculum, if not an official or required one.

Leaving the mathematics issues to the mathematicians, I'll summarize "three key shifts" for the English language arts.

First is a "regular practice with complex texts and their academic language" (Common, "Key Shifts"). There's a call for "a staircase of increasing complexity" as students move through the curriculum. *Complexity* is defined specifically (and controversially) in terms of both systematic vocabulary development and a formula applied to readings. Those readings are to come from multiple domains, including "classic myths and stories from around the world, foundational U.S. documents, seminal works of American literature, and the writings of Shakespeare." Good old Will gets his customary personal shout-out, but the whole of literature is but one township among a wide terrain of texts. To illustrate the variety they have in mind, the Common Core developers list example readings, which, though traditional, have served to stir conservative concerns that the federal government is taking over schools. I've been a local witness to this from the ongoing school protests in Jefferson County, Colorado, a district that begins a mile from my house in Denver (Healy). Tea Party school board member Julie Williams called for the Advanced Placement history curriculum to focus on patriotism, respect for authority, and free enterprise and to exclude materials that encourage civil disorder. In a Facebook press release, Williams blamed the Common Core generally, and David Coleman specifically, for almost every ill up to the demise of American democracy.

The second key shift in the language arts is to privilege "reading, writing, and speaking grounded in evidence from texts, both literary and informational" (Common, "Key Shifts"). The standards framework asserts that previously student writing had mainly been from student experience and opinion. It calls instead for a "significant shift from current practice" to "text-dependent" or source-based writing to prepare students for "the demands of college, career, and life," those demands dependent on informing and persuading. In distinguishing among narrative, informative, and argumentative writing, the standards reproduce the venerable mix of modes and aims. In devaluing narrative, they seem to miss the role that story now plays in corporate and popular culture.

The third key shift is "building knowledge through content-rich nonfiction" (Common, "Key Shifts"). This is perhaps the most unsettling thing for school English teachers, since it calls for a fifty-fifty split between literary and informational reading in grades K–5 (Common, "Key Shifts" no. 3). The standards for grades 6–12 call for more literary nonfiction. But as a scholar of the personal essay, memoir, new journalism, and so on, I'm pretty sure the Common Core definition of literary nonfiction differs from mine. There's not much Joan Didion or John McPhee or Alison Bechdel or David Foster Wallace, even if the kinds of works listed in the eleventh-grade suggestions, for example, are plenty worthwhile:

"Speech to the Second Virginia Convention" by Patrick Henry (1775)
"Farewell Address" by George Washington (1796)
"Gettysburg Address" by Abraham Lincoln (1863)
"State of the Union Address" by Franklin Delano Roosevelt (1941)
"Letter from Birmingham Jail" by Martin Luther King Jr. (1964)
"Hope, Despair and Memory" by Elie Wiesel (1997)
Common Sense by Thomas Paine (1776)
Walden by Henry David Thoreau (1854)
"Society and Solitude" by Ralph Waldo Emerson (1857)
"The Fallacy of Success" by G. K. Chesterton (1909)
Black Boy by Richard Wright (1945)
"Politics and the English Language" by George Orwell (1946)
"Take the Tortillas Out of Your Poetry" by Rudolfo Anaya (1995)

Perhaps as radical as the call for nonfiction and literary nonfiction, the standards require literacy instruction in history / social studies, science, and technical subjects. When the overview explains that "reading, writing, speaking, and listening should span the school day from K–12 as integral parts of every subject," longtime advocates for writing across the curriculum should well cheer.

The attention to disciplinary conventions themselves, though, ought to give us some pause. From early discourse-community theory, to genre theory, to activity-systems theory, some composition scholars have established rationales for teaching students, averring that different traditions and epistemologies manifest different rhetorical practices. But starting to teach that in high school is problematic.

To summarize, the Common Core framework shifts literature away from the center it held in the twentieth-century English curriculum. It places broader emphasis on nonfiction and reading and writing across the curriculum. College faculty in rhetoric and composition will likely be heartened that the standards require more source-based informative, analytic, and argumentative writing. They'll appreciate how the new emphases challenge arhetorical formalistic pedagogies like the five-paragraph theme, the modes of discourse, and power writing. Now, I have great faith (and not a little despair) that entrepreneurial publishers and consultants will devise and package new formulae and attendant curricula. Similarly, if I were a textbook publisher, I'd chain a whole warehouse full of MFA graduates to workbenches, having them put out nonfiction books carefully deployed on a stairway of complexity that steps kids through grade levels.

Those concerns aside, the short forecast for a more complex view of reading and writing looks sunny, especially to those of us long concerned about how narrow and literature-centric the school English curriculum has sometimes been. Yet I see clouds on the horizon.

Concerns about the Standards

In many respects, the current outcry has derived not from the standards per se but from their implementation. There's been a sort of perfect storm in which aspects of No Child Left Behind, Race to the Top, and the Common Core Standards have intermingled. No Child Left Behind, that bipartisan effort whose progressive critics tend to forget was cosponsored by Ted Kennedy, mandates that one component of teacher evaluations must be student growth. Students don't show learning, teachers don't get raises. Student performance is to be a fraction of the total rating, 15 to 50 percent. The rub, of course, is how to measure that performance, and the Common Core Standards assessments offer a new mechanism for doing so, even as the No Child Left Behind mechanisms have been vexed. In recent years the Department of Education administration has been granting waivers that allow states to delay implementing those measures, and even the Brookings Institution sees little hope for accountability through testing when 80 percent of schools are failing to meet targets (Whitehurst). As a result, implementation timetables have been all over the place, with a relatively small number of states now having teacher accountability determined at least partly by student test performance and most other states moving to delay the step (Ujifusa).

Complicating the relationship among standards, testing, and accountability was the Race to the Top program, which awarded $5 billion through state-by-state competitions that preferred states that had adopted the Common Core Standards. As a result, the wagon of state-developed standards (again, the impetus came from the governors, not from the feds) got hitched to the federal star. And while the standards don't mandate federal assessment or data collection, we're in a neoliberal age of accountability, big data, and school report cards. If you're going to have standards, the gravitational force is toward measuring them, especially since No Child Left Behind measurements have already been in place for a decade. (Of course, the Every Student Succeeds Act replaced No Child Left Behind in 2015.)

The standards are national, of course, and there's a reasonable argument that students and teachers in a global, mobile, and digital society might benefit from national assessments. Arguments for local or state control over education often result in censorship and narrow curricula that restrict both certain texts and subject matters and also certain approaches, such as those calling for critical analysis.

If it's logical to have national standards and national assessments, then it's also sensible to pursue economies of scale in developing them. Thus, two testing consortia have emerged: Partnership for the Assessment of Readiness for College and Careers, or PARCC, and the Smarter Balanced Assessment Consortium. States can join one or neither, and by February 2015, ten had states joined PARCC and eighteen, Smarter Balanced, totaling twenty-eight members (Education Week, "National"). Or, to look at this from another perspective, twenty-two states have opted out of the consortia, some of them having once belonged but having since opted out, substantially for political reasons in conservative states. Indiana, for example, is creating the Indiana Academic Standards, Arizona the AzMERIT exam for college and career readiness (Arizona Department of Education). States' relationships to the testing consortia continue to be in flux, even as the tests themselves get modified.

For example, in May 2015, in response to feedback from schools and states, PARCC consolidated two testing windows into one, reduced the number of testing units, and shortened the testing time (PARCC Communications Team).

The federal government provided PARCC and Smarter Balanced some $360 million by October 2014 to develop assessments. The consortia, in turn, farmed out aspects of this process to over a dozen vendors, the largest among them McGraw-Hill, ETS, and Pearson (Education Week, "Money"). McGraw-Hill, for example, spent $72 million developing test items, processes, and scoring guides, including $358,000 for "achievement level descriptors." (The rubric for argumentative writing, grades 6–11, for example, has two parts: purpose/organization and evidence/elaboration, each of them a four-point rubric that generally matches what rhetoric and composition teachers might develop.) The tests themselves have both direct and indirect measures, which means that students have to write. All of that writing needs scoring, but human scoring on this scale is complex and expensive, so the testing consortia dream of computer scoring. Smarter Balanced, for example, declares that it "will capitalize on the precision and efficiency of computer adaptive testing (CAT) for both the mandatory summative assessment and the optional interim assessments" (Smarter Balanced, "Computer"). While machine scoring may work for certain kinds of indirect assessments, it's clear that we're a long way from valid and reliable assessments of writing itself (NCTE, *NCTE Position*).

Now, at this point I'm fairly much in the procedural and political weeds and fairly far from the standards themselves. However, it's crucial to understand how goals get transmuted by testing imperatives and logistics. The thought experiment of "what if the standards actually work" depends on understanding both the standards themselves and how they're implemented through testing.

A Glance at First-Grade and Eleventh-Grade Language Arts Standards

Consider, briefly, the standards developed for two grade levels. First graders are expected to perform three kinds of writing: opinionated, informative, and narrative. They are to

> write opinion pieces in which they introduce the topic or name the book they are writing about, state an opinion, supply a reason for the opinion, and provide some sense of closure.
>
> Write informative/explanatory tests in which they name a topic, supply some facts about the topic, and provide some sense of closure.

> Write narratives in which they recount two or more appropriately sequenced events, include some details regarding what happened, use temporal words to signal event order, and provide some sense of closure. (Common, "English Language . . . Grade 1")

Whereas previous generations of six-year-olds were mainly honing the muscle control to hold pencils to make words and sentences, first graders under the core are expected to produce texts that have explicit formal features.

Moreover, in producing those texts, first graders also should display a particular set of skills (with an apology to Liam Neeson's *Taken* character), some of them "with guidance and support from adults." They're to respond to peer suggestions and add details to strengthen writing and to use "various digital tools" for production and publication, including collaboratively. They're to participate in research projects, for example, exploring a "number of 'how-to' books on a given topic and use them to write a sequence of instructions" (Common, "English Language . . . Grade 1"). The reading and writing abilities here are a far cry from the kinds of writing my own first grader did twenty-five years ago, writing his name and telling a story about a dog, with scrawled drawings, little research, and scantly a sense of closure.

If school is serious business for first graders under the standards, it's hugely so for grades 11–12. Given explicit skepticism about experiential or personal writing, it's not surprising to see a strong focus on claims, valid reasoning, and substantive evidence. Seventeen-year-olds are to

> introduce precise, knowledgeable claims(s), establish the significance of the claims, distinguish the claims(s) from alternate or opposing claims, and create an organization that logically sequences claim(s), counterclaims, reasons, and evidence.
>
> Develop claim(s) and counterclaims fairly and thoroughly, supplying the most relevant evidence for each while pointing out the strengths and imitations of both in a manner that anticipates the audience's knowledge level, concerns, values, and possible biases. (Common, "English . . . Grade 11–12")

There's much to like here, obviously, and who mightn't hope that students who came to college and careers—not to mention citizens who show up in voting booths—would have these abilities? There's even a whiff of audience awareness and rhetorical situation. Any of my reservations are more with the nature of adolescent development, the way their logical reasoning develops in concert with other language abilities (including creative ones absent here), and the stark contrast between the idealized discourse posited in the standards and what currently manifests as civic discourse in our society, where evidence-free assertion and pathos reign. We risk breeding cynics. In fairness to the standards' authors, the privileging of rational argument could be seen as a welcome attempt to improve civic discourse—which may account for strong opposition from political ideologues.

The grades 11–12 standards also foreground disciplinary knowledge and conventions, presenting separate achievement levels for non-English disciplines in both the sciences and social sciences. In this sort of writing, students are to "convey a knowledgeable stance in a style that responds to the discipline and its context as well as to the expertise of likely readers" (Common, "English . . . Grade 11–12.2.D"). They're to use formatting, graphics, and multimedia in ways that advance disciplinary conventions.

The standards for grades 11–12 are clearly ambitious, and they obviously differ from recent practice. Most scholars of rhetoric and composition can applaud writing for rhetorical purposes, with content and features suggested by audiences and disciplines rather than being starkly universal. That said, the standards raise two questions. First, how developmentally appropriate are they? Can most students learn to perform these kinds of tasks, especially in the psychological and social milieu of high school and adolescent development? Are students in a reasonable zone of proximal development—students in hardscrabble Cicero, Illinois, as well as in white-collar Cherry Creek, Colorado? Second, how do these standards construct writing for students and schools? No doubt, informing and arguing from source materials is important in school, work, and civic life. But to what extent is this kind of writing singularly important? What about kinds of writing largely occluded from the standards: experiential, creative, exploratory, or social rather than informational or agonistic?

Testing Approaches and the Construction of Writing

To explore these questions, consider the tests themselves, which have been created for digital environments. In January 2014, I was a member of the NCTE Government Relations Committee that met with three gracious Department of Education officials in Washington, DC, to discuss Common Core assessments. It was clear that the lack of technology in many schools was making large-scale testing difficult, a realization that seemed to be surprising to both governmental and testing industry officials. It's hard to judge student proficiency with digital tools when those tools are absent or present but scantly.

One sample exam available on the PARCC website in 2014 gives students a complex task. They're to read the Declaration of Independence and a speech by Patrick Henry, and they're to watch a video on the Declaration produced by "the Kettering Foundation, a non-profit organization that focuses on helping democracy in the United States as the Founders intended it to function" (PARCC, "Declaration"). Ultimately, students have to write an analytic essay. But before that stage come prewriting and invention tasks.

For example, in one series of screens, the two written documents scroll in a left window, and multiple-choice questions scroll in the right. Similarly, a video link gets activated in the left window alongside a question in the right.

Next, students answer a question that generates a proto-thesis: "Based on the three sources, what was a major assumption shared by Thomas Jefferson and Patrick Henry?" (One possible choice, for example, is "Even though they are right to rebel against Britain, the colonists should be equally aware of the dangers of conflict within the colonies themselves.") That done, they're to choose two quotations that support their answer, from a multiple-choice answer. All along through this process, they can toggle among the three documents available in the left window. This presumes a certain facility with managing information and composing in digital environments, no doubt important skills but not ones routinely taught in traditional literacy classrooms. One can imagine the test makers either as naïve or, more charitably, as goading pedagogical changes—expectantly accompanied by generous budgets for professional development and access to technology.

Finally comes the full writing task: "Write an essay in which you explore the government's purpose presented in the sources. . . . Remember to use evidence from all three sources to support your ideas" (PARCC, "Declaration"). The students get a word-processing window with a few formatting tools. This is a rigorous task, especially for on-demand timed writing. Consider, most mundanely, what students might take *explore* to mean—and whether the scoring rubric for this task would, in fact, reward or penalize an exploratory essay primarily trying to weigh possibilities. I suspect the rubric would penalize. Beyond skills in close reading, analysis, and synthesis, the task presumes navigational skills in a technological environment and competence for retaining, somehow, information gleaned online and deploying it, without the advantage of note making, in a well-structured writing produced on the spot. Most students don't receive that kind of instruction today. Again, the standards may be goading and aspirational, for the good, but we need to take great care that testing results not be used to judge students, teachers, and schools, in our characteristically reductive "report card" manner.

Both PARCC and Smarter Balanced sample exams feature a range of other kinds of scaffolded writing. A second task asks students to read four sources about financial matters, in preparation for writing a report to a school board. The board has invited students to submit arguments about whether the district should require a financial literacy course for graduation (Smarter Balanced, "Mandatory"). In one text box, students are asked to declare which of the four sources is most useful and to support and justify their claim with two details. In a second box, they're to "paraphrase information from Source #1 that refutes information from Source #2 without plagiarizing." In a third activity, students are presented with a number of claims and are told to "click on the boxes to show the claims each supports."

In a third type of task, students are given a scenario: write an argument about a proposed curfew law (Smarter Balanced, "High School Student"). They're given curious materials: another student's letter to the president of the city council, along with that student's drafting notes. One sentence in the letter is underlined, and the student test taker is told, "Using the underlined topic sentence, complete the second paragraph" of the original letter. In this exercise, then, students are told to inhabit a particular position and use ideas and arguments presented to them, rather than formulate their own argument and evidence.

In a fourth type of scaffolded strategy, students are ultimately asked to write a comparative analysis of two passages about Daedalus, one from Ovid, one from Anne Sexton (PARCC, "Daedalus"). They're directed to select a theme for their analysis from a multiple-choice list of options. On a subsequent screen, they encounter a number of quotations from the two poems, with an empty box labeled "Evidence" at the bottom. They're told to "select two quotations that provide additional evidence to support the answer to Part A. Drag and drop your answers into the box labeled Evidence."

In a final example, students are assigned the role of someone writing a report on how to get rid of hiccups (Smarter Balanced, "Hiccups"). They encounter a screen with titles, brief synopses, and brief identifying information about four online sources, ranging from scholarly to governmental to popular ones. Directions ask them to evaluate each source in terms of its relevance and credibility.

Now, I could continue providing examples from the sample exams, but I trust that several things stand out. First, all the writing situations rely on reading and analyzing texts, almost always more than one, often reasonably complex works. Second, students must perform a series of tasks that structure invention for them: selecting from given potential claims, matching available evidence to those claims, judging efficacy, and so on. On the one hand, these operations serve to provide pathways for students in a testing situation that's likely stressful, for texts that are more likely foreign. On the other, they provide multiple data points, especially for computer-adaptive testing, which can feed into "objective" reading and writing scores. Third, students do this analysis in a screen environment that requires operations like scrolling through windows, toggling between windows, dragging and dropping, typing in windows only to use that typing later in drafting, formatting texts in progress, and so on.

There's no doubt that these future writers will encounter workplace and academic situations that require using chunks of data in constrained amounts of time. And there's some elegance in providing materials and structuring operations for those constraints; we may even perceive an element of fairness here, though it's clear that rhetorical situations such as writing to city councils and school boards will be considerably more familiar, friendly, and plausible to students from some environments than to students from others. Further, there's an element of game playing to the tasks, something akin to a first-person play-through of an environment, albeit in a graphically poorer playerscape. This isn't necessarily a bad thing, mind you, but we shouldn't mistake the view of reading and writing promoted in these online assessments as constituting *all* reading and writing.

A constricted sense of writing is, finally, the aspect of the Common Core Standards that most merits scrutiny. Especially as enacted through the current assessments—those assessments portending curricula and activities—writing acquires a rather technical cast, as a series of manipulations and operations performed sequentially, even algorithmically, to reach an acceptable end state. To be fair, certain facets of writing largely *can* be cast as technical manipulation, and it's surely true that much school writing has been less authentic or purpose driven than it has been formalistic performance.

Ultimately questionable is less the value of this kind of writing per se (it clearly has value) than its partiality in terms of the kind of writing to be valued and practiced in schools. Readiness for college and careers, part of the PARCC acronym, is a worthy literacy goal for high school graduates, surely, and enmeshed in that goal are skills useful for other kinds of literate goals, including civic readiness and personal development and social and relational writing. However, there are purposes for reading and writing beyond the instrumental ones pursued at desks, however vital those might be. And, in fact, those "soft" skills, with their attention to the kinds of reasons one may choose to read and write rather than be obliged to do so, might be crucial to the kind of deep engagement that fosters true practical facility. This is especially true for our very youngest students, those first graders whose early encounters with language form their far horizons. One might take solace in knowing that nothing in the Common Core Standards precludes a broader view of writing; the standards describe a minimum, not a maximum. Yet there is ever a truth that what is tested is what gets taught.

What If It Works?

Let me circle back to my central question: What if the Common Core Standards succeed? That is, what if a large percentage of American students actually achieve the language arts standards, through a radical transformation of school writing instruction? I've noted several potential benefits above, including more emphasis on reading and analyzing nonfiction, more emphasis on argument and exposition for readers other than English teachers, and recognition that writing conventions vary across disciplines. All desirable.

These gains come at two main costs. First, we might expect students to enter college with an even larger disconnect between the kinds of writing they experience in school and the kinds they encounter in the whole of life beyond. In the wider civic and social sphere, attitudes and actions are shaped by canons of rhetoric beyond logos, for better and, often, for worse. We might hope the Common Core will eventually raise all of civic discourse to a higher plane of assertion and evidence. That's a grand years-long cultural project that I, too, devoutly desire, but meanwhile I worry that we foster cynicism if students don't explore connections between the idealized world of discourse presented in the Common Core and social media's wild frontier—and, these days, it's fairly much all social media. Partly this comes through having students read, discuss, and critique day-to-day texts, even the tawdry world of comments, alongside canonical or model texts. Partly this comes through having students practice forms of writing that are shorter, messier, and more dependent on clever phrasings or evocations of memes. It's both/and. That is, even as we teach complex, well-formed, aspirational reading and writing, we need to help students understand the connections and disconnections between those utopic forms and often dystopic actual practice. Failing to do that makes school reading and writing precisely only that.

Such disconnections are widened in light of a second concern, namely, the near absence of personal, interpersonal, and social reasons that people write. Consider a few types of reading, researching, and writing that are hard to find sanctioned in the standards. A student interviews a grandparent about his experiences in Vietnam or the first Gulf War and writes an oral history of what he learns. A student in an online gaming community shares some experiences and advice to others playing the game; she writes to explain the game and why it matters to readers who aren't in that community. A student takes a trip and takes pictures—a trip across town or across the state—and renders the experience in words and images so that he can remember and share it. A student reads a series of novels or magazines because she likes science fiction or vampires or hiking or hip-hop, some of it well enough to write fan fiction or contribute to discussion boards. (For example, I participate in an online forum of Colorado mountain climbers, 14ers.com, whose members post extensive trip reports, often describing in detail their routes and equipment.) These are reading and writing activities well beyond college and career readiness. No doubt, lots of reading and writing are obliged: we do them because there are consequences if we don't or because we want to effect some change. But lots of reading and writing are self-sponsored, done for autotelic reasons or purely to make connections with others. If there is some truth in Mr. Coleman's complaint about overreliance on experience, there's a larger truth that experience and personal interest motivate deep engagements with language. Tests may do that in the short term, provided that external costs and rewards are sufficiently compelling, but an education that fails to acknowledge the role and sources of individual engagement—the attractions of pursuing and rendering kinds of experience, however seemingly impractical—is an education that chooses to abjure a path to sustaining itself.

Higher education routinely invokes interdisciplinarity or transdisciplinarity, in which intellectual activity is organized, not around acquiring bodies of knowledge or mastering epistemological or discursive practices characteristic of individual fields, but around problems or questions. Mainly this stems from knowing that most complex problems have multiple facets drawing on different fields. Increasing literacy proficiencies among second-generation Mexican American students in Denver public schools, for example, involves aspects of sociology, history, politics, and economics as well as education and language development theory. Historically, this movement has given rise to institutes or research centers or projects. What's interesting in the current environment is that interdisciplinarity increasingly seems a vital aspect of undergraduate education. Mainly this stems from recognizing that while prizing disciplinary distinction may be useful for graduate studies and convenient for organizing curricula, it's less justified as an organizing principle for undergraduate studies—and less compelling to students who see compelling issues manifesting not as disciplinary questions but as "actual" or "life" questions. Now, no doubt, disciplinary histories and frameworks bring kinds of deep attention to those issues, and no doubt, college professors bring to readings of student writing tacit expectations for "good writing" that are steeped in disciplinary traditions. There's a role, then, for the Common Core Standards in raising consciousness of how disciplinarity shapes texts. And there's a role, too, for pedagogies that use well-structured questions, accompanied by well-chosen texts, to teach strategies and skills.

However, true college readiness these days requires a broader array of dispositions and aptitudes. The *Framework for Success in Postsecondary Writing*, developed by the Council of Writing Program Administrators, NCTE, and the National Writing Project, discusses eight vital habits of mind needed for college writing: curiosity, openness, engagement, creativity, persistence, responsibility, flexibility, and metacognition. Several of these are probably fostered by the Common Core, yet the standards foster what rather might be seen as technical facilities, with the goal of having students perform certain reading and writing operations that can be tested and certified. There's an implicit faith that the habits noted in the *Framework* will come later or, somehow, in parallel. However, by failing to attend to language practices as they actually and fully exist in the public sphere, and by failing to account for how personal and interpersonal motivations for reading and writing complexly interact with schooling imperatives, the Common Core may inadvertently complicate certain aspects of college writing, trading curiosity for competence, openness for closure, engagement for performance.

I'm overstating the oppositions to make a point, of course, and reiterate (for at least the third time) that there's much to like about the standards. We just need to recognize that the very aim of these standards—college and career readiness—casts college in a particular, somewhat tidy, way and serves particular ends. And we need to remember that people use language for all sorts of purposes beyond the important ones of earning degrees or acquiring jobs and that those other uses are key to the kind of deep and sustained engagements with reading and writing that surely the Common Core's creators desire.

Works Cited

Arizona Department of Education. "Assessment: AzMERIT." *Arizona Department of Education*, www.azed.gov. Accessed 1 Mar. 2015.

Coleman, David. "Bringing the Common Core to Life." Albany, New York. 28 Apr. 2011. *New York State Education Department*, usny.nysed.gov/rttt/docs /bringingthecommoncoretolife/fulltranscript.pdf. Speech.

Common Core State Standards Initiative. "About the Standards." *Common Core State Standards Initiative*, www.corestandards.org.

———. "English Language Arts Standards>>Writing>>Grade 1." *Common Core State Standards Initiative*, 2015, www.corestandards.org. Accessed 25 Sept. 2014.

———. "English Language Arts Standards>>Writing>>Grade 11–12." *Common Core State Standards Initiative*, 2015, www.corestandards.org. Accessed 25 Sept. 2014.

———. "Key Shifts in English Language Arts." *Common Core State Standards Initiative*, 2015, www.corestandards.org. Accessed 25 Sept. 2014.

———. "Myths vs. Facts." *Common Core State Standards Initiative*, 2018, www.corestandards.org/about-the-standards/myths-vs-facts/.

Council of Writing Program Administrators, National Council of Teachers of English, National Writing Project. *Framework for Success in Postsecondary Writing*. 2011. PDF.

Education Week. "Money Flowing for Common Core Assessments." *Education Week*, 1 Oct. 2014, www.edweek.org/ew/section/multimedia/consortia-dollars-for-common-core-testing.html?cmp=ENL-EU-NEWS1.

———. "The National K-12 Testing Landscape." *Education Week*, www.edweek.org/ew/section/multimedia/map-the-national-k-12-testing-landscape.html. Accessed 20 Feb. 2015.

Healy, Jack. "In Colorado, a Student Counterprotest to an Anti-Protest Curriculum." *The New York Times*, 23 Sept. 2014, www.nytimes.com/2014/09/24/us/in-colorado-a-student-counterprotest-to-an-anti-protest-curriculum.html?_r=0.

Murphy, Tim. "How a Bipartisan Education Reform Effort Became the Biggest Conservative Bogeyman since Obamacare: Inside the Mammoth Backlash to Common Core." *Mother Jones*, Sept./Oct. 2014, www.motherjones.com/politics/2014/09/common-core-education-reform-backlash-obamacare/.

National Commission on Writing. *The Neglected "R": The Need for a Writing Revolution*. College Entrance Examination Board, 2003.

———. *Writing: A Ticket to Work . . . or a Ticket Out?* College Entrance Examination Board, 2004.

National Council of Teachers of English. *NCTE Position on Machine Scoring*. 15 Apr. 2013, www.ncte.org/library/NCTEFiles/Resources/Positions/MachineScoring_single_page.pdf.

National Governors Association. "Common Core State Standards Available for Comment." Press release, 21 Sept. 2009. Web.

PARCC. ["Daedalus and Icarus"] "PARCC Sample Set HS ELA>7 of 23." *Partnership for the Assessment of College Readiness and Careers*, epat-parcc.testnav.com/client/index.html#getitem/10186. Accessed 28 Feb. 2015.

———. ["Declaration of Independence"] "Grade 11 ELA/Literacy/Session 1/11 of 17." *Partnership for the Assessment of Readiness for College and Careers*. Web. Accessed 28 Feb. 2015.

PARCC Communications Team. "PARCC States Vote to Shorten Test Time and Simplify Test Administration." *Partnership for the Assessment of Readiness for College and Careers*, 28 May 2015, www.coreeducationllc.com/blog2/parcc-states-vote-to-shorten-test-time-and-simplify-test-administration/.

Smarter Balanced Assessment Consortium. "Computer Adaptive Testing." *Smarter Balanced Assessment Consortium*, www.smarterbalanced.org. Accessed 25 Feb. 2015.

———. ["Hiccups"] ELA HS Training Test (5 out of 6). *Smarter Balanced Assessment Consortium*, www.smarterbalanced.org. Accessed 25 Feb. 2015.

———. "A High School Student Is Writing a Letter." ELA HS Training Test (2 out of 6). *Smarter Balanced Assessment Consortium*, www.smarterbalanced.org. Accessed 25 Feb. 2015.

———. "Mandatory Financial Literacy Classes Argumentative Performance Task." G11 Performance Test. *Smarter Balanced Assessment Consortium*, www.smarterbalanced.org. Accessed 25 Feb. 2015.

Southern Poverty Law Center. *Public Schools in the Crosshairs: Far-Right Propaganda and the Common Core State Standards*. 30 Apr. 2014. Web.

Ujifusa, Andrew. "Teacher, School Accountability Systems Shaken Up." *Education Week*, 12 June 2014. Web.

Whitehurst, Grover J. "The Future of Test-Based Accountability." The Brown Center Chalkboard, *The Brookings Institution*, 10 July 2014. Web.

Williams, Julie. "Press Release: Re: AP U.S. History Curriculum Committee." *Facebook*, 23 Sept. 2014, www.facebook.com/julieforjeffco/posts/1514318528813222. Accessed 28 Sept. 2014.

Afterword:
The Not-So-Simple Truth of English B

Jessica Enoch and Scott Wible

> *The instructor said,*
> *Go home and write*
> *a page tonight.*
> *And let that page come out of you—*
> *Then, it will be true.*
> *I wonder if it's that simple?*
>
> —Langston Hughes, "Theme for English B" (1951)

In "Theme for English B," Langston Hughes writes of the false simplicity of English B, a course that is for him riddled with race and power and inflected too by concerns of age and location. In Hughes's poem, the student writer reflects on his assignment for English B and his relationship with his white teacher, and he challenges the simple truth his teacher asks for: "it's not easy to know what's true for you or me." Hughes's words signal the complexity of English B, the required writing course that is the subject of this collection, and his resistance to see the theme as a simple truth is one the contributors to this volume embrace, meditate on, and compound. While much about higher education in general and the composition course in particular has changed in the more than fifty years since Hughes penned his poem, the complexity of the course—the truth that this course is anything but simple—has stayed the same and will most certainly continue to do so.

In this volume of fifteen chapters, leading scholars in the field of rhetoric and composition speak to the complexity of teaching composition, all articulating their own ideas regarding the issues that teachers of academic and professional writing face as we look to the future and anticipate the complexities before us: What is our not-so-simple truth now, and what will it be in 2020, 2045, or even 2100? What should we be concerned with and why? Building from the exciting presentations and conversations at the 2014 Maryland Conference on Academic and Professional Writing, this collection's contributors think through these questions and supply us not with simple answers but with robust responses that add depth and dimension to Hughes's poem. In their prognostications these scholars helpfully and provocatively move from the theoretical and political to the everyday and the practical, offering big ideas, hard questions, and immediate pedagogical possibilities. This collection is a marker of the moment—what were our concerns for English B in 2014?—and a guide for our futures—what should we care about and how should we proceed in teaching composition?

Hughes first and primarily points to the ways that race animates the composition classroom. He writes, "So will my page be colored that I write? / Being me, it will not be white." As Hughes could have certainly anticipated, this connection between race and writing instruction still does and will continue to require critical attention and pedagogical invention for scholars of rhetoric and composition. In her chapter, Staci Perryman-Clark highlights the range of complex challenges that language educators and writing program administrators (WPAs) will need to address in earnest over the coming decades, from creating and sustaining infrastructure that supports historically marginalized students in writing courses to examining and, where necessary, challenging teacher attitudes toward marked and unmarked racial, ethnic, and linguistic differences in student writing. The demographics of higher education will most certainly continue to change as the United States itself becomes more racially, ethnically, and linguistically diverse. Perryman-Clark highlights the problems inherent in so many white, monolingual teachers perceiving themselves to be the last line of defense in preserving English language "standards," but as Langston Hughes urges us to see, teachers share in the truth that students live and write:

> *You are white—*
> *yet a part of me, as I am a part of you.*
> *That's American.*
> *Sometimes perhaps you don't want to be a part of me.*
> *Nor do I often want to be a part of you.*
> *But we are, that's true!*

Perryman-Clark's essay, like Hughes's poem, challenges us to see that writing programs need to do more than simply find ways to help linguistic minority students reproduce the seemingly universal, immutable language standards of English B. Instead, the challenge for writing programs moving forward will be to reimagine the learning aims of English B, to reenvision English B as a space where students learn to critically examine and purposefully draw from their entire repertoire of language resources to produce academically substantive, publicly engaged compositions. The presence of language diversity in our classrooms is a part of our rhetoric and composition discipline, and we need to radically re-see the aims of the course as a result.

Suresh Canagarajah's chapter helps us in this work of reimagining what the linguistic aims of English B could and should be. The common assumption in much first-year writing curriculum and pedagogy, he explains, is that the learning goals should be for all students to work toward native-like competence in a single language. Canagarajah instead sees opportunity for redefining this educational norm given the increasing presence in writing classrooms of students who possess competencies in more than one language or dialect. Canagarajah contends that rhetoric and writing pedagogy needs to take translingual speakers as the educational aspiration and find ways to help all students learn how to draw productively from multiple languages and dialects in order to produce rhetorically effective texts in any given rhetorical situation. Canagarajah urges us to support educational policies that promote students' learning of multiple languages and to create activities, assignments, and assessments that help students learn to marshal the language resources they possess and find creative ways to draw on them to communicate in any situation, rather than evaluating students' language use in terms of how closely it adheres to a single, static norm. Canagarajah importantly calls then for a new orientation toward the writing course, a new conceptualization of both what and how students can articulate their truths through their compositions.

In the same way that Perryman-Clark and Canagarajah compel teachers to reimagine the aims of the writing classroom and to create space for students to draw on a broad range of language resources to meet those learning goals, Anne Frances Wysocki asks teachers to heighten their awareness of how they respond to the aesthetic dimensions of texts. One thinks of the student writer in "Theme for English B," who expresses a belief in the ways that his multisensory life experiences shape who he is:

> *But I guess I'm what*
> *I feel and see and hear, Harlem, I hear you:*
> *hear you, hear me—we two—you, me, talk on this page.*

Wysocki asks us to think about our responses, not simply to the language varieties students use to tell their "truth," but also to the sensory experiences they respond to and try to compose on the page. She prompts us to examine the judgments we pass about the aesthetics of texts, and she presents heuristics to help us think about how we might cultivate writing from our students that draws on the full range of meaning-making resources, not simply logical progressions of arguments but also textual and visual representations of sensory experiences. The aims for Hughes's student writer and for Wysocki's imagined composition classroom, then, are the same: for teachers and students to understand how guiding assumptions about traditional composition pedagogy privilege particular ways of seeing or hearing or feeling texts and to expand the range of rhetorical and aesthetic resources they draw on to achieve their aims in any print or digital text they compose.

In addition to examining the ways that assumptions about race, language, and sensory experience shape how both students compose and teachers respond to writing, the contributors to this collection echo Hughes by attending to the effects of location on teaching and learning. The student writer in "Theme for English B" acknowledges the physical and cultural distances between community and school when he writes,

> *I am colored, twenty-two, born in Winston-Salem.*
> *I went to school there, then here*
> *to this college on the hill above Harlem.*

Location, Hughes reminds us, matters to students, particularly for those who cross not only physical but also significant social distances as they travel to schools up "on the hill." The contributors to this collection confirm this importance, asserting that location matters just as much to teachers and administrators as it does to students. Kelly Ritter writes in her chapter that all "pedagogy is local," and we must understand that while any overarching pedagogical theory may seem enticing or any global concern may seem important, these ideas and concerns only gain relevance when they are placed in the particular circumstances of our institutions and the specificity of our classrooms. We must translate the theoretical to the practical and explore how our everyday practice speaks back to overarching ideas and scholarly conversations. In other words, and as Cheryl Glenn reminds us, we must identify and understand our administrative and pedagogical priorities from within our own mesosystems.

Chapters in this collection from past and present WPAs—Ritter, Glenn, Paul Sawyer, Theresa Redd, Jeanne Fahnestock, Bob Coogan, Jane Donawerth, and Molly Scanlon—are deeply important because these WPAs speak from their institutional contexts to articulate the complexity of their location-based "truths." Their investigations of their mesosystems are important not just because they allow readers to see how programs came to be and where they are going but also because they offer us moments of contemplation, translation, and (dis)identification, enabling us to ask questions: How did we come to do what we do? Is this happening where I am? How might I translate this practice within my institutional context? How does my pedagogical and administrative experience speak back to this idea or concern?

Because the 2014 conference was held at and hosted by the University of Maryland, it makes perfect sense that two of the presentations and two of the chapters collected here would be dedicated to the academic and professional writing programs at that university, the same programs that the writers of this afterword now direct. The storied histories authored by Coogan, Donawerth, and Scanlon and by Fahnestock in conversation with the other contributors reveal the changes that writing programs undergo, and they helpfully articulate the surprising and unsurprising catalysts to these changes. Their narrative reflections make clear how programmatic change at Maryland was inspired by national conversations and large-scale federal imperatives such as the civil rights movement and the Morrill Act as well as what Fahnestock identifies as individual "dissatisfactions" with the way things were at Maryland, highlighting the point that the monumental, the mundane, and the idiosyncratic are all consequential to writing program administration.

In these two Maryland chapters, however, it is also clear that throughout the history of writing instruction at this institution and in the face of all these changes, a "curricular mainstay," as Coogan, Donawerth, and Scanlon write, has been "preserved": rhetoric. Fahnestock confirms this pedagogical consistency over decades of instruction, and she even moves beyond it to pinpoint what this consistency actually looks like to her. Fahnestock writes that a rhetorically based course must focus on the following features: "an attention to audience" and "rhetorical invention"; an "awareness that all discourse is 'making a case'"; and an appreciation for both the "idea that cases are made with different degrees of certainty or confidence" as well as the idea that language offers the writer a "set of positive options at the word, sentence, passage, and whole-text level." Thus, reflecting on the past and seeing both consistency and change, Coogan, Donawerth, Scanlon, and Fahnestock offer readers a moment to consider their own institutional pasts, to identify curricular mainstays that define their programs, and especially to identify what if anything they might add or take away from the elements of the rhetorical core that Fahnestock describes.

To be sure, all of the chapters in conversation with those by Coogan, Donawerth, Scanlon, and Fahnestock make one thing clear: there is nothing simple about running a writing program. In their descriptions, Glenn, Redd, Ritter, and Sawyer articulate their pedagogical investments: the connections among reading and writing, genre awareness, process and revision, civic participation, research, inquiry, digital composing, and rhetoric. And each

WPA's description of the many moving parts to coordinate reveals an almost mind-boggling array of concerns: multilevel courses and assignment sequences; learning outcomes and assessments; pedagogical training, practicums, and seminars for new and experienced instructors; living-learning communities and service learning; and collaboration with writing centers and other university programs, just to name a few. Most important to these WPAs, though, is the concern for the student. Both Ritter and Glenn ask a version of the same question, seeing it as the impetus behind all of their work: How do (or should) our writing courses best serve our particular students?

Echoing the concerns of Canagarajah and Perryman-Clark summarized above, Ritter describes her unique situation at Illinois in terms of its student body:

> since 2008, the undergraduate international student population at Illinois has more than doubled—rising from 2,232 to 4,996 students. Much of this growth has been driven by rising numbers of Chinese international undergraduates, whose population grew from 258 in 2008 to 2,588 in 2014.

Ritter interprets the significance of these numbers, explaining, "Currently, Illinois has the largest international student population of any public university in the United States." Illinois's student population, then, consists of residents from the "Chicagoland" area, a "smaller percentage of minority and underrepresented students," and also the growing numbers of international students. The critical question for her, then, is what is the best writing program and what are the best writing courses that meet the needs of *all* these students? And her answer, for sure, will be different from those of Glenn, Sawyer, Fahnestock, and Redd because the needs of the students who are learning and writing in our classrooms push us to make our pedagogies local.

While Ritter calls attention to students in their particularity and diversity, Sawyer and Glenn underscore additional concerns that complicate administrative work. Sawyer writes of the grim situation for adjunct teachers at schools all over the country, as well as the often-uncomfortable position of the WPA as one of lower management. Within institutional hierarchy, the WPA might have some control over the courses she administers, but she works ultimately at the behest (and sometimes the whim) of the upper management of department chairs, deans, provosts, and presidents. Sawyer points specifically to the deplorable situation that recently emerged at Arizona State in which upper managers "raised the teaching load for writing faculty from four-and-five to five-and-five without a raise in salary." Sawyer's account thus prompts these questions: As WPAs, how do we speak back to these untenable demands on teaching? How do we protect our instructors and create environments for them in which they not only survive (passably) but thrive? Sawyer "reminds us that high standards of employment correlate directly with high standards of teaching." To us, these seem to be words to live by.

Glenn's chapter similarly zeroes in on the limited powers of WPAs. Her rich review of feminist perspectives on writing program administration by scholars such as Sally Barr-Ebest, Lynn Bloom, Theresa Enos, Susan Miller, Eileen Schell, Krista Ratcliffe, and Rebecca Rickly signals the collective voices who have investigated and experienced WPA work as feminized—something seen as "women's work" and thereby "neither serious, rigorous, nor intellectual." But it is Glenn's discomfort with this designation and her own exploration of how "the powers of rhetoric and feminism" can invigorate administrative work that bring poignancy to this feminist scholarship. In her chapter she tells of the *feminist* (not feminine) style of negotiation she used when Penn State's faculty senate began planning for an ill-devised and slapdash review of the university's general education curriculum that would certainly affect the students and instructors in Glenn's writing program. Leveraging the feminist rhetorical practice of invitational questioning, Glenn was able to "slow down the process" of the review to allow for more collaboration and dialogue. It was her expertise in feminist rhetoric that enabled her to cultivate "productive, mutually empowered conversations about the general education review." Thus, while the powers of the WPA are limited—surely we cannot do it all—there are modes of intervention we can deploy to make our programs and institutional environments more humane.

To be sure, each of the contributors in section II points to the importance of knowing our place, of studying the contours of our mesosystem for the ends of effective writing program administration, because we know that teaching in Harlem is indeed different from teaching at Illinois. But many of the authors in this collection also make it clear that we are not alone and should not see our administrative work in isolation. Fahnestock especially underscores this point when she relates the productive relationships that administrators at Maryland built with Edward Corbett at Ohio State and James Kinneavy at Texas. Redd celebrates similar connections when she tells of the visit that WPAs from Howard made to Cornell. At Cornell, Redd writes, Howard WPAs "had seen WAC in action, and they liked what they had seen." Redd further elaborates on this idea of relationship building when she investigates another problematic kind of isolation: not an institutional isolation but a programmatic one.

Redd argues that a key concern for WPAs should be to resist the idea that writing instruction is "boxed" into English departments or writing programs. Instead, Redd makes clear, we should think creatively about breaking out of the box to create cultures of writing that permeate our campuses. Redd's inventive Writing Matters Campaign reveals how, in addition to supplying writing advice and pedagogical instruction to instructors at Howard, there was also a need for instructors to attest to the level of rigor they were expecting from students. That is, if students know that teachers beyond the composition classroom are setting high expectations for their writing, they will, in turn, take their own writing more seriously. Redd emphasizes the importance of these expectations, especially for the African American students at Howard. She explains that students at her institution may have encountered educational environments in which the bar was not set high for them even though they could reach it. Redd's argument is a persuasive one: we all must create programs that meet our students where they are, but we must also push them to move forward, to stretch and challenge themselves as writers.

Kathleen Blake Yancey similarly seeks to destabilize our seemingly routine approaches to reading and responding to student writing. While she values the work that has been done to create more holistic assessments, like portfolio assessments, that consider students' abilities to produce effective textual responses to a range of rhetorical situations, she also points to the missed opportunities for teaching and learning that result when assessment remains wholly in the purview of the teacher. Like the student in "Theme for English B," who tries to create a space for his personal experiences and insights within the academy, the students described in Yancey's chapter would compose reflective pieces and impact statements that enable them to engage effectively in the process of assessment. Yancey's vision for assessment even applies pressure to the notion of English B as a gateway course, for she illustrates how students can build on the electronic portfolios they create in first-year writing to document, examine, and articulate how they learn to use writing across a range of disciplines, genres, and curricular and extracurricular sites throughout their college careers.

Yancey sees promise in electronic portfolios, as does Doug Eyman in digital, networked composing, for challenging ingrained, habituated ways that teachers often teach and read student writing. Yancey and Eyman call on teachers to attend to the full spectrum of textual and design elements that writers can draw on to craft their texts to achieve their purpose in a specific physical or virtual context. Indeed, Eyman warns that our conditioned responses to defining *composition* and *text* must necessarily evolve with the constant changes in digital and networked technologies. He promotes broader thinking about the kinds of projects we put before students and the frames we provide to help them understand not only the what and how of composing in digital genres but also the why. Specifically, Eyman asks us to envision a holistic conception of composition, one that complicates the notion that the writing course should focus solely on helping students learn how to develop the "content" of their texts. Like both Yancey and Wysocki, Eyman hopes to foster a more robust conception of text, one that asks students to move beyond the *theme* for English B and think instead about every single detail of a text as an available means: the words, of course, and the rhetorical appeals that give them shape and meaning; but also the graphical elements of texts, such as typeface, size, and alignment; page layout and document design; and visual images used either as stand-alone or integrated elements of the text.

Eyman not only calls for expanded notions of what counts as integral elements of print and digital texts but also, like James Paul Gee and Doug Hesse, asks us to imagine composition courses as spaces for creating a broader range of writing experiences for students. For one, Eyman wants students to gain opportunities for composing procedural rhetorics, such as computer coding, that create the infrastructure that makes the construction and circulation of digital texts possible. Eyman theoretically frames these computer-coding assignments as a means to teach students about the rhetorical resources that computer coding makes available to writers and designers in the digital realm.

Hesse, meanwhile, reflects on the constraints and opportunities at play in the Common Core Standards designed for college and career readiness, acknowledging the positive components of the regime that calls on students to produce more writing, to learn how to work more effectively with sources, and to compose with a clearer sense of audience, purpose, and genre than the traditional five-paragraph essay assignment encourages them to do. At the same time, Hesse contends that the Common Core, like so many school-based literacy programs before it, once again threatens to restrict students' conception of writing. To be sure, the increased attention to informational and persuasive writing is significant, and the focus on disciplinary conventions, while perhaps too advanced a concept for high school writers as Hesse suggests, nevertheless orients students to the types of discursive concerns they will need to manage in college or careers. "But to what extent," Hesse asks, "is this kind of writing singularly important?" He asks us to create a future for students with an even broader range of writing experiences for them, ones that allow for experimentation with creative, experiential, exploratory, and social writing. This pedagogy would create opportunities for students to discover personal motivations and interests for writing rather than performing only those writing activities for assignments they are obliged to do for college and career readiness.

James Paul Gee's contribution to this collection provides the theoretical foundation for a writing pedagogy that allows for this type of experimentation and play with writing. His exemplification (through the popular video and card game *Yu-Gi-Oh!*) of what he calls Situated Learning Systems helps us to reimagine how we can design our writing classrooms to enhance student motivation and deepen their learning. Teachers need to create opportunities for students to "live in and muck around in the world that the language of the text is about," and we need to be flexible and adaptable with our formal instruction to follow so that we're helping students build on their emerging understanding of how language works in any particular genre or discourse community. Both Hesse's and Gee's essays, then, remind us of the student in "Theme for English B," who encounters a traditional, rather generic writing assignment but then experiments with it, discovers how particular styles of writing do or do not enable him to articulate what he "feels and sees and hears," and uses those insights as a means to analyze and reshape accepted ideas of what counts as sources of academic and cultural knowledge.

Like Yancey, Eyman, Gee, and Hesse, Charles Bazerman and James Herrick both anticipate the ways that technology will change our experiences as teachers and students. Assessing the expansive kinds of work that emergent computer technology can now accomplish, Bazerman asks, "While machines will come to do what machines do best, humans must reallocate their attention and skills to do what humans do best." We ask with trepidation what's left for us humans given the long (long) list of what computers can now do: voice recognition and transcription; bibliographic technologies and citation searches; distant reading and idea-mapping software; and the list goes on. Bazerman does not push us too deeply into despair, however, as he identifies the kinds of work that humans are especially adept at and should continue to cultivate: higher-level decision making, collaboration, audience analysis and relations, intuition, imagination, inspiration. The challenge for writing teachers and administrators will be to continue to develop our curriculum and our pedagogy so that students are gaining, refining, and excelling in these areas where computers cannot (yet) go. And those pedagogies, Bazerman makes clear, should be quite familiar to us, as he emboldens us to recommit ourselves to practices of revision, rhetoric, collaboration, and inspiration.

James Herrick makes a similar return to the familiar in the face of great technological change. Assessing the "world [his] digital-native students inhabited," Herrick writes that he sought an "approach to teaching written and spoken expression and criticism that was flexible, practical, and grounded in real-world language practices." He soon discovered, though, that he did not need to "reinvent the discursive wheel when such a well-tested conveyance was so close at hand and so many skilled pilots could be so readily summoned." The answer was one he already knew. Now more than ever, students need to cultivate their skills in managing evidence, interpreting abbreviated expression such as tweets, unpacking enthymemes, cultivating skills of inquiry, and engaging publics of interest. In short, students need to expand and deepen their understanding of rhetoric. Thus Herrick's conclusions are much the same as those of Bazerman, who asserts that "the future will be much the same, only different. Or perhaps the future will be different, but much the same."

This collection in its entirety confirms Hughes's point: English B, its students, its teachers, its administrators and administration, and of course its pedagogies are not simple. Given all the complexity we face, we could indeed feel quite anxious about where we're going and how we'll get there, and how we can, along the way, treat our students, teachers, and administrators fairly, while honoring our students' past and present identities and preparing them for the unknown future. On the other hand, though, we can feel somewhat secure in noting that many of the contributors here circle back to the ancient origins of rhetorical study. Even in the face of great technological change, rhetorical study and expertise might once again be our most effective means of teaching students to write their varied truths into the fields of academic and professional writing.

These essays compel us to listen carefully to our students—who they are, where they come from, how they make meaning in the world, where they want to go, and who they want to be in their future professional, civic, and personal lives—as we chart a future direction for the fields of academic and professional writing. As he closes "Theme for English B," Hughes writes, "As I learn from you, / I guess you learn from me." This perspective—we learn from our students as much or even more than they learn from us—must guide our research, teaching, and writing program administration. We should create spaces in our classes for students to explore their truths and to examine the languages, genres, modes, and technologies they use to compose these truths, and we must continue to expand the range of contexts for which students write and the types of writing they use to write themselves into those contexts. Rhetorical theory presents a solid foundation for our field to teach and to analyze this writing, but we must continue to revisit and revise these theories through the lenses of different cultures, languages, technologies, and bodily experiences. Moreover, the future of academic and professional writing demands that we create spaces for teachers and WPAs from different cultures, nationalities, and language backgrounds to compose and share their own truths, as well, and to reshape perspectives on academic and professional writing as they do. The contributors to this volume, like Hughes, show us how context matters, how experience matters, how identities formed through language, culture, technology, and institutions of schooling matter in the practice of writing. Our future for the teaching of writing will move forward productively and capaciously only if we continue to identify, attend to, and wrestle with the varied and complex truths of English B.

Contributors Index

Contributors

CHARLES BAZERMAN, distinguished professor of education, University of California, Santa Barbara, is interested in the social dynamics of writing, rhetorical theory, and the rhetoric of knowledge production and use. His books include *A Rhetoric of Literate Action, A Theory of Literate Action, The Languages of Edison's Light, Shaping Written Knowledge, The Informed Writer,* and *Handbook of Research on Writing.*

SURESH CANAGARAJAH is the Edwin Erle Sparks Professor and the director of the Migration Studies Project at Pennsylvania State University. He teaches world Englishes, teacher development, and postcolonial studies in the Departments of English and Applied Linguistics. His early education and teaching were in war-torn Jaffna, Sri Lanka. He later taught for the City University of New York.

ROBERT COOGAN is a professor emeritus in the Department of English at the University of Maryland. When asked to reform the freshman writing course, which had been a mix of literature and writing, Professor Coogan, like the Renaissance humanists, went back to the fount, relying on adaptations of the rhetorical theory of Aristotle. To improve the teaching of incoming teaching assistants, they had a graduate course, but, more important, they were assigned mentors to guide them through their first year. The mentors created the esprit de corps found in joyful teamwork.

JANE DONAWERTH, a professor emerita of English at the University of Maryland, was the director of introductory writing from 1984 to 1988 and the director of writing programs from 2012 to 2014. She is the editor of *Rhetorical Theory by Women before 1900: An Anthology* (2002) and the author of *Conversational Rhetoric: The Rise and Fall of a Women's Tradition, 1600–1900* (2012).

JESSICA ENOCH is an associate professor of English and the director of academic writing at the University of Maryland. Her publications include *Refiguring Rhetorical Education: Women Teaching African American, Native American, and Chicana/o Students, 1865–1911* (2008) and *Domestic Occupations: Spatial Rhetorics and Women's Work* (2019). She has also coedited a collection with Dana Anderson, *Burke in the Archives: Using the Past to Transform the Future of Burkean Studies* (2012), and the critical anthology she coedited with Cristina Ramírez, *Mestiza Rhetorics: An Anthology of Mexicana Activism in the Spanish-Language Press, 1887–1922,* is set for publication in 2020.

DOUGLAS EYMAN teaches courses in digital rhetoric, technical and scientific communication, and professional writing at George Mason University, where he directs the PhD program in writing and rhetoric. Eyman is the senior editor and publisher of *Kairos: A Journal of Rhetoric, Technology, and Pedagogy,* an online journal that has been publishing peer-reviewed scholarship on computers and writing since 1996.

JEANNE FAHNESTOCK is a professor emeritus in the Department of English, University of Maryland. She is the author of *Rhetorical Style* (2011; RSA Outstanding Book, 2012) and *Rhetorical Figures in Science* (1999, 2002) and is a coauthor of *A Rhetoric of Argument* (third edition, 2004). Fahnestock is a fellow of the Rhetoric Society of America and a distinguished scholar (2014) of the International Society for the Study of Argumentation.

JAMES PAUL GEE is the Mary Lou Fulton Presidential Professor of Literacy Studies and Regents' Professor at Arizona State University. He is a member of the National Academy of Education and has published widely in journals in linguistics, literacy studies, learning science, and education.

CHERYL GLENN is the University Distinguished Professor of English and Women's Studies at Pennsylvania State University. Her scholarly publications include *Rhetorical Feminism and This Thing Called Hope*; *Rhetoric Retold: Regendering the Tradition from Antiquity through the Renaissance*; *Unspoken: A Rhetoric of Silence*; edited collections; textbooks; and articles and chapters. In 2015 she was awarded an honorary doctorate from Örebro (Sweden) University. In 2008 she served as chair of the Conference on College Composition and Communication.

JAMES A. HERRICK is the Guy Vander Jagt Professor of Communication at Hope College in Holland, Michigan, where he has taught since 1984. Herrick is the author of *The History and Theory of Rhetoric*, *The Radical Rhetoric of the English Deists*, and *Visions of Technological Transcendence*. He teaches critical thinking, the history of rhetoric, and the rhetoric of technology.

DOUG HESSE is a professor of English and the executive director of writing at the University of Denver, where he was named Distinguished Scholar. Past president of the National Council of Teachers of English, he previously chaired the Conference on College Composition and Communication and was president of the Council of Writing Program Administrators. Hesse's seventy-plus articles and four coauthored books focus on creative nonfiction and professional issues in writing and English studies. His PhD is from Iowa. He's a hiker, a photographer, and a singer.

SHIRLEY WILSON LOGAN is a professor emerita of English at the University of Maryland and the author of *With Pen and Voice*, *We Are Coming*, *Liberating Language*, and a number of essays on nineteenth-century African Americans' rhetorical practices. She is a coeditor of the Southern Illinois University Press series Studies in Rhetorics and Feminisms and has held several professional leadership positions, including chair of the Conference on College Composition and Communication.

STACI PERRYMAN-CLARK is an associate professor of English and associate dean of the Lee Honors College at Western Michigan University. She is the author of *Afrocentric Teacher-Research: Rethinking Appropriateness and Inclusion* (2013) and (with David E. Kirkland and Austin Jackson) a coeditor of *Students' Right to Their Own Language: A Critical Sourcebook* (2014).

TERESA M. REDD is a professor emerita at Howard University, where she recently directed the Writing across the Curriculum Program and the Center for Excellence in

Teaching, Learning, and Assessment. She has published two books, *Revelations: An Anthology of Expository Essays by and about Blacks* and (with Karen Schuster Webb) *Introduction to African American English: What a Writing Teacher Should Know.*

KELLY RITTER is the associate dean for curricula and academic policy in the College of Liberal Arts and Sciences and a former director of the Undergraduate Rhetoric Program (2013–17) at the University of Illinois at Urbana-Champaign. Her latest book is *Reframing the Subject: Postwar Instructional Film and Class-Conscious Literacies* (2015). With Melissa Ianetta, she is a coeditor of the forthcoming *Landmark Essays in Writing Program Administration* (2018).

PAUL SAWYER received his PhD from Columbia University and is a professor of English at Cornell University. His book *Ruskin's Poetic Argument* (1985) explores the work of the Victorian art critic, radical economist, and polymath. He was the director of the Knight Institute for Writing in the Disciplines (2006–16) and is a founder of the Cornell Prison Education Program.

MOLLY J. SCANLON is an associate professor in the Department of Writing and Communication in the College of Arts, Humanities, and Social Sciences at Nova Southeastern University. She teaches graduate and undergraduate courses in composition, writing studies, and visual rhetoric. Her research interests include visual rhetoric, comics studies, and identity construction, and she has publications in *Composition Studies*, *Reflections*, and *ImageTexT*.

WAYNE H. SLATER is an associate professor in the Department of Teaching and Learning, Policy and Leadership at the University of Maryland. His research is grounded in cognitive psychology, linguistics, and rhetoric. His publications have appeared in the *Journal of Educational Psychology*, *Reading Research Quarterly*, *Research in the Teaching of English*, and *Reading and Writing Quarterly*. He has been awarded five Elementary and Secondary Education Act Title II grants.

SCOTT WIBLE is an associate professor of English and the director of the Professional Writing Program at the University of Maryland. His research examines rhetorics of public policy and innovation pedagogies in professional writing. His book *Shaping Language Policy in the U.S.: The Role of Composition Studies* (2013) won the Conference on College Composition and Communication's 2014 Advancement of Knowledge Award.

ANNE FRANCES WYSOCKI is an associate professor of English, emerita, at the University of Wisconsin–Milwaukee, where she taught undergraduate and graduate courses in written, visual, and digital rhetorics.

KATHLEEN BLAKE YANCEY, the Kellogg W. Hunt Professor of English and a Distinguished Research Professor at Florida State University (FSU), has served as president or chair of the Council of Writing Program Administrators, the Conference on College Composition and Communication (CCCC), and the National Council of Teachers of English. A past editor of *College Composition and Communication*, she has authored, edited, or coedited fourteen books and over one hundred articles and book chapters. Her awards include FSU's Graduate Mentor Award and CCCC's Exemplar Award.

Index